The Blohm & Voss Ha 222 was ordered by Deutsche Lufthansa as a passenger flying boat for the North Atlantic route.

Blohm & Voss BV 222 "Wiking"

Rudolf Höfling

Introduction

The aircraft produced by the Dornier Metallbauten GmbH of Friedrichshafen on Lake Constance in the 1920s and 30s provided the foundation for the successful development of German flying boats. The company also achieved a prominent international reputation in this specialized aviation field, especially with the Dornier Wal flying boat and the extraordinary twelve-engined Dornier Do X.

It was thus scarcely imaginable that an outsider would be able to penetrate into Dornier's domain, yet that is what happened. Thanks to its chief designer Dr.-Ing. Richard Vogt, beginning in the mid-1930s the Hamburger Flugzeugbau GmbH, a subsidiary of the famous Blohm & Voss shipyard, gradually challenged Dornier's claim of being the sole provider of flying boats to Lufthansa and the Luftwaffe. In addition to the successful three-engined Blohm & Voss BV 138, the Hamburger Flugzeugbau GmbH and its chief designer Dr.-Ing. Richard Vogt secured their place in the history of German seaplane development with two more extraordinary designs the Blohm & Voss BV 222 Wiking and the Blohm & Voss BV 238. These two six-engined giants owed their creation to two entirely different customer requirements specifications. While the elegant Blohm & Voss BV 222 was designed to meet a requirement of the airline Lufthansa, from the beginning the even bigger BV 238 was conceived purely for military use.

The Hamburger Flugzeugbau GmbH (HFB) was founded in June 1933 by the brothers Rudolf and Walther Blohm, owners of the Blohm & Voss shipyard also located in Hamburg. Faced with a worldwide economic crisis, the brothers hoped that the new production branch would provide an additional mainstay, and they were counting on government contracts following the ascension to power in Germany of the National-Socialists at the end of January 1933. In fact, contracts for the construction under license of components of the Junkers W.34 and Ju 52/3m were very soon issued, which also led to the first expansion of the company's facilities. The company did not restrict itself to copying the products of others, also making space for its own developments.

The company's first chief designer was Reinhold Mewes, who came from the Ernst Heinkel Flugzeugwerke GmbH and developed the Blohm & Voss Ha 135 biplane trainer, which made its first flight in the spring of 1934. The obsolescent steel tube construction of the fuselage and the type's wooden wings were not well received, however, and the Blohm brothers soon began looking around for a new technical director. They very soon found one, for Dr.-Ing. Richard Vogt seemed to be the right man for the advertised position. Vogt had served as a

1

The P.54 design by Dr.-Ing. Richard Vogt ultimately led to the Blohm & Voss BV 222.

pilot in the First World War and in 1917 had been attached to Zeppelin Flugzeugbau in Lindau. There he gained his first experience in aircraft design and met Claude Dornier, who was then a close associate of Count Ferdinand Adolf August Heinrich von Zeppelin. After the war and various other activities, Dornier hired him for his own aircraft company in 1922. Dr.-Ing. Richard Vogt accepted an offer to go to Japan to oversee the licensed production of Dornier aircraft. The initially limited stay ultimately turned into ten years. After the conclusion of his actual activity with the Kawasaki Dockyard Ltd. in Kobe, he continued his work in Japan and developed several of his own aircraft. For Richard Vogt the offer from Blohm & Voss was so enticing and lucrative that in 1934, after returning to Germany, he accepted the position offered him as chief designer for the Hamburger Flugzeugbau GmbH. No one then suspected that he would earn a worldwide reputation in the design of flying boats. During his eleven years as chief designer, Vogt designed a total of 13 aircraft for Blohm & Voss.

His first flying boat project, the Blohm & Voss Ha 138 (later BV 138), was a difficult learning experience for Vogt. Extensive redesign work and a resulting lengthy development period were required before the flying boat entered service. His second flying boat design, however, was an immediate success. The aircraft was the Blohm & Voss Ha 222, which together with the BV 238 represented the apex of German flying boat development. It should be mentioned here that in 1938 the original subsidiary, the Hamburger Flugzeugbau GmbH, was integrated into the parent firm, as a result of which all subsequent aircraft designs were designated "BV" instead of "Ha".

The German airline Deutsche Lufthansa hoped to establish a passenger service on the lucrative North Atlantic route to the USA with the Blohm & Voss Ha 222. With the Ha 222, Richard Vogt presented a flying boat which in many respects set new standards. This was true not just of its aerodynamic and hydrodynamic design, for many new ideas were also incorporated into the giant's control system.

By the time the flying boat took off on its maiden flight in 1940, the Second World War had already broken out. Because of an error in reasoning by its decision-makers, most of the aircraft operated by the Luftwaffe were designed for short- or medium-range operations. This included the transport fleet, made up mainly of Junkers Ju 52/3m trimotors. Early in the war Great Britain's intention to occupy northern Norway to secure important sources of raw materials became apparent, resulting in the first long-range operations by the Luftwaffe in the Second World War. Lacking suitable aircraft, the Luftwaffe was forced to turn to the few four-engined passenger aircraft of the airline Lufthansa. The Junkers Ju 90, Focke-Wulf Fw 200 and Dornier Do 26 saw action, as well as the totally obsolete Junkers G 38, which had been the world's largest aircraft at the start of the 1930s. It was therefore predictable that the Blohm & Voss BV 222 would be called into use as a freight and troop transport as soon as possible. But the German navy also wanted this long-range aircraft for use as a strategic reconnaissance aircraft over the Atlantic.

The German aviation industry lacked three things in the late 1930s: first it lacked sufficient trained personnel for the construction of aircraft, second certain raw materials were in short supply, and third the German aero-engine industry was as yet incapable of delivering powerful engines in quantity. These shortcomings affected the production of large aircraft in particular and consequently the Blohm & Voss BV 222. As well, the Luftwaffe could not employ its few long-range aircraft in concentration, for it constantly had to balance the requirements for transports and long-range reconnaissance aircraft. The ultimate result of this was that in the coming years of the war it had neither sufficient numbers of long-range transports nor strategic reconnaissance aircraft. The BV 222 was also affected by this dichotomy. Initially used as an unarmed transport, it later took on the role of long-range reconnaissance aircraft, which obviously required the addition of defensive armament. Although the BV 222 proved a success in this new role, as the war went on more aircraft and their crews were lost as the Allies' material superiority made the air war over Europe and North Africa increasingly difficult.

As a result of the previously-mentioned shortcomings in the German aviation industry and the shifting of priorities caused by the war, resulting in increased production of fighter aircraft, plus the frequent Allied air raids on production sites in Northern Germany, just thirty Blohm & Voss BV 222 Wiking flying boats were completed. One can only speculate as to the type's chances as a civilian aircraft. They would probably have been limited, however, because of the appearance of a new generation of land-based aircraft like the Junkers Ju 90 and Focke-Wulf Fw 200 that became available in the second half of the 1930s.

Beginning in the mid-1930s, a trend developed mainly in the USA and Great Britain of using large flying boats as civil transports on long overseas routes. In the United States, Boeing and Martin developed suitable aircraft for Pan American Airways. These "Clippers" saw service mainly in the Atlantic and Pacific regions. In Great Britain, Short Brothers Ltd. assumed a leading position in flying boat design. In addition to its most well-known military flying boat of the Second World War the Short Sunderland approximately ten years earlier the company had developed a line of large flying boats for civilian use: the Short Empire Class. These four-engined flying boats mainly served the air routes to North Africa and India operated by Imperial Airways. At the end of 1936 Lufthansa issued a speci-

The P.45 Trans-Oceanic Airplane Project and its Competing Designs

fication for a trans-Atlantic flying boat for its own North Atlantic routes to the three aircraft manufacturers Heinkel, Dornier and Hamburger Flugzeugbau.

Dr.-Ing. Richard Vogt was one of the most inventive, adventurous and innovative aircraft designers in Germany. His six-engined trans-oceanic flying boat project would justify this reputation. A technical description of this design, which was designated the P.45, has survived.

The "Takeoff Boat:

The most significant disadvantage of a flying boat compared to a land-based aircraft is its "boat fuselage", which must be made significantly wider than that of a land-based aircraft and is thus aerodynamically unsatisfactory on account of its high drag. Dr.-Ing. Richard Vogt therefore set himself the task of finding a solution to this problem. So he came up with the idea of the so-called "takeoff boat". The idea was that an aerodynamically-efficient aircraft should be placed on this "takeoff boat", which itself possessed optimal hydrodynamic characteristics. After the aircraft achieved takeoff speed, several guides, which had been kept under tension by springs or compressed air the whole time, would lift the aircraft from the "takeoff boat" and in this way dramatically reduce its takeoff run. Dr.-Ing. Richard Vogt developed a second boat as an alternative to the "takeoff boat", which could be attached rigidly to the fuselage of the aircraft. The point of this was to make it possible to eliminate the time-consuming process of mounting the flying boat on its "takeoff boat" on land for maintenance test flights or flights of short duration.

The Blohm & Voss P.45 Project

The P.45 Project originated in 1937 and had a flat "boat bottom", as it was intended to lift off from the water with the help of a "takeoff boat". In flight, therefore, the P.45 resembled a conventional land-based aircraft with a retractable undercarriage and similar aerodynamics. After landing on the water, a pair of auxiliary floats housed in the wings similar to those later realized in the design of the Blohm & Voss BV 222would stabilize the flying boat. It was envisaged that the aircraft would be powered by six Junkers Jumo 205 diesel engines each producing 500 hp (368 kW). The Junkers Jumo 205 diesel engine was built in various versions which, beginning with the early variants, delivered 500 hp, while the final version, the Jumo 205 D which powered the Blohm & Voss BV 138, produced 880 hp. The engines were to be accessible in flight by means of a crawlway in the main spar, a special design feature that was a hallmark of Blohm & Voss seaplanes and preferred by Dr.-Ing. Richard Vogt. The tubular spar increased the bending and torsional rigidity of the wings and tail surfaces. Such tubular spars also provided space for various tanks.

Though the design's calculated theoretical performance was convincing, it also had to provide a high level of comfort for its passengers. With a width of between 2.5 and 2.6 meters, the fuselage was divided into an upper and a lower deck, with stairs in the rear of the flying boat linking the two decks. For long-range operation, an interior layout for 24 passengers was envisaged. This provided them with comfortable armchairs in the lower deck and beds in the upper deck. The P.45 was so roomy that it offered a twelve cubic meter freight compartment plus bedrooms and a toilet for the crew. The P.45 project had a calculated wingspan of 46 meters, the same as the later Blohm & Voss BV 222, a length of 33.3 meters, height of 6.6 meters and a wing area of 310 square meters. It was estimated that the P.45 would achieve a range of 8000 kilometers and a maximum speed of 410 kph, with an empty weight of 26 000 kg and a maximum takeoff weight of 51 000 kg.

On 19 September 1937 Deutsche Lufthansa placed an order for three examples of the Blohm & Voss BV 222. Seen here is the first prototype.

LUFTWAFFE AIRCRAFT Profile

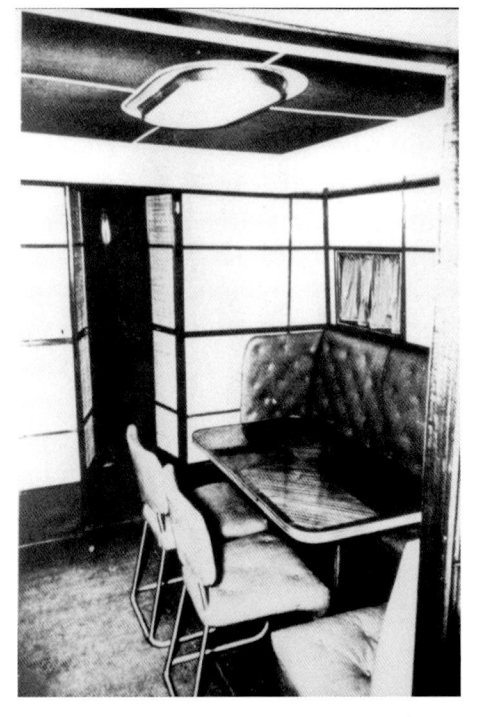

A 12.5-ton crane was used to place the tubular spar on the fuselage of the BV 222. The tubular spar was a characteristic design feature of Dr.-Ing. Richard Vogt.

On 16 July 1940 representatives of Deutsche Lufthansa inspected the mockup of the Blohm & Voss BV 222. The photo shows part of the interior layout of the flying boat.

Blohm & Voss P.54 Project

Despite all the advantages of Dr.-Ing. Richard Vogt's design, Deutsche Lufthansa could not warm to his P.45 Project. In the opinion of Lufthansa management, the use of a separate "takeoff boat" would require a very complex and expensive ground organization, which would not exist, especially outside Germany, or would be extremely difficult and cost intensive to establish. Dr.-Ing. Richard Vogt therefore returned to his drawing board and developed the P.54, a new and more conventional flying boat which would ultimately enter production as the Blohm & Voss BV 222.

The Dornier Do 20 Project

Two years before the start of the Blohm & Voss P.45 project, the Dornier Metallbauten GmbH launched its own project, the Dornier Do 20. The design was largely based on the Dornier Do X "Airship" of 1929. For example the aircraft employed a typical design feature of Dornier flying boats the stub wing. The straight wing with round wingtips and the "boat hull" of the Do X were adopted almost unchanged for the Do 20. The flying boat's power plants were new, however. The Do 20 was to have been powered by eight diesel engines (probably Junkers Jumo 205s) each producing 800 hp (588kW) and buried entirely within the aircraft's thick wing. Each pair of engines was to have driven a single propeller by means of a shaft, giving the impression of a four-engined flying boat. The Do 20 was designed for a range of 4000-5000 km and would have carried between 12 and 16 passengers on long-distance flights. In addition to sleeping cabins, the aircraft would have offered generously-proportioned sitting rooms. A different arrangement was envisioned for shorter flights with a seating capacity of up to 60 passengers.

In 1936 a model of the Dornier Do 20 was shown to the public for the first time at the International Aviation Exhibition in Stockholm. With a wingspan of 49 meters, a length of 40 meters, a height of 9.5 meters and a wing area of 450 square meters, the dimensions of the Do 20 were roughly similar to those of the Dornier Do X. The Dornier Do 20's empty weight was supposed to have been 29500 kg and its takeoff weight 50000 kg. The flying boat's cruising speed was about 290 kph, a significant improvement over the 175 kph of the Do X, which was six years older. Deutsche Lufthansa nevertheless decided against the project and selected the P.54 design by the Hamburger Flugzeugbau GmbH.

The Heinkel He 120 Project

Conceived in 1938, for a time the Heinkel He 120 flying boat, a shoulder-wing monoplane with gull wing, was seen as a rival of the Blohm & Voss BV 222. If one compares the size and performance of these "rivals", however, doubts arise. Conceived for North and South Atlantic routes, the He 120 was to have been powered by four Junkers Jumo 205 diesel engines each delivering 800 hp (588 kW), which was to have given the aircraft a cruising speed of 265 kph and a maximum speed of 380 kph. The manufacturer also guaranteed a range of 6900 km. Depending on how it was configured, the flying boat could have carried up to 25 passengers. With less power than the BV 222 and a lower empty weight of about 29000 kga wingspan of 35 meters, a length of 28 meters and a wing area of 170 square meters the smaller He 120 was not comparable to the Blohm & Voss BV 222 and thus could not be seen as a direct rival design. Deutsche Lufthansa was also interested in the aircraft whose streamlined shape was spoiled by fixed stabilizing floats and a full-scale mockup was built. By order of the State Ministry of Aviation (RLM), however, the Ernst Heinkel Flugzeugwerke GmbH was forced to abandon development of the flying boat in 1939.

Because of delays associated with the war, the roll-out of the Blohm & Voss BV 222 bearing the civilian registration D-ANTE did not take place until the end of August 1940.

After creation of the State Ministry of Aviation (Reichsluftfahrtministerium) in 1934, state influence in the aviation industry grew to an unprecedented scope and shape. The first measures included the creation of new production facilities as well as the removal of disliked personalities from the existing firms. Dr. Hugo Junkers was a prime example, as was Ing. Heinrich Lübbe of the Arado Flugzeugwerke GmbH. Both men were forced to leave their companies under pressure from the RLM. The State Ministry of Aviation's influence was already such that no new aircraft could be built without the approval of the ministry. As a state-controlled organization, therefore, Deutsche Lufthansa's ability to make independent decisions was severely limited, especially as the procurement and development of new military and civilian aircraft was usually financed and therefore controlled by the State Ministry of Aviation.

The Heinkel He 220 Project

The RLM's order to halt development work on the He 120 did not stop Ernst Heinkel from proceeding with a follow-on project, the He 220 flying boat. In addition to the crew, the aircraft was designed to carry between 32 and 48 passengers and had split retractable stabilizing floats. Like the smaller He 120, the He 220 was a shoulder wing monoplane with a gull wing. Its four power plants, each producing 3,500 hp (2,572 kW), were supposed to give the 70 000 kg flying boat a cruising speed of 430 kph. Maximum range was 8700 kilometers. Not surprisingly, the He 220 proceeded no farther than the drawing board, for it was also rejected by the State Ministry of Aviation in favor of the BV 222. It is also questionable where Heinkel was supposed to obtain such powerful engines for his flying boat design in the early 1940s.

Before opening the story of the Blohm & Voss BV 222, a few remarks concerning procurement policy in the Third Reich:

Lufthansa had ordered the BV 222 as a trans-Atlantic flying boat (first flight by the V1 on 7/9/1940). During the war the Luftwaffe's largest flying boat served as a transport and reconnaissance aircraft.

The Blohm & Voss BV 222 Flying Boat – The Beginnings

Within the ranks of the big national aviation companies, Deutsche Lufthansa (until 1 April 1934 written as Luft Hansa), which had been founded in 1926, played an important role. The Company had earned a solid reputation through numerous pioneering feats and long-distance flights in the 1920s and 1930s. This was not enough to succeed in international competition, however. Establishing profitable air routes had to be foremost amongst the company's activities.

In the mid-1930s, aviation was still in its infancy worldwide. Improvements in flight safety and expansion of the ground organization were slow in coming. The United States of America, Great Britain, France and Italy therefore decided on the flying boat as a long-range aircraft. Its greatest advantage was its ability to land on just about any river, lake or sea in the event of mechanical trouble. The flying boat had no need of expensive runways for takeoff and landing. In keeping with this international trend, at the end of 1936 Deutsche Lufthansa requested the construction of a long-range flying boat for use mainly on the North Atlantic routes. The new aircraft had to combine high performance and great comfort, for only thus could Deutsche Lufthansa compete against its American, British, French and Italian rivals, who also offered air service to North America.

The demands that Deutsche Lufthansa placed on this new aircraft were clearly outlined. The new flying boat had to be able to cross the Atlantic non-stop, carrying at least 24 passengers by day and 16 by night. It also had to offer a level of comfort comparable to that of the big Zeppelin airships then in operation. The desired crossing time was 20 hours, consequently the aircraft's cruising speed had to be about 330 kph. In contrast, the German airship LZ-129 Hindenburg's fastest crossing from Lakehurst to Frankfurt required 42 hours and 53 minutes. After delivering its list of requirements, Deutsche Lufthansa left the designers a free hand in conceiving their design. One of the airline's demands was that flying boat be able to continue its flight after the loss of an engine.

During flight trials the V1 was given the standard German camouflage scheme for maritime aircraft and the military code CC+EQ.

After submission of the extensive technical description of the P.54 project on 31 May 1937, on 19 September Deutsche Lufthansa issued a contract for the completion of three prototypes under the type designation Blohm & Voss Ha 222. In the mid-1930s the Hamburger Flugzeugbau GmbH had two establishments located in Steinwerder and Wenzendorf. The final assembly lines

The Blohm & Voss BV 222 V1 on its beaching apparatus.

The transport missions flown by the V1 also served as flight trials.

for the Blohm & Voss Ha 138 and Ha 139 seaplanes were located in Steinwerder. It soon became apparent, however, that this was a less than favorable location for the construction of seaplanes. From the factory in Steinwerder the completed aircraft had to be placed by crane onto a pontoon or the Blohm & Voss shipyard's recovery vessel, the B & V Kranich, and then transported to the Elbe River for testing. This was a very laborious procedure, and at the turn of the year 1935-36 plans were being finalized for the construction of a new facility. Called Development Factory Finkenwerder, it was built to serve land- and water-based aircraft. Work on the new factory began in early 1937 and was completed in March 1940. Two 9000-meter-square hangars formed the core of the new factory in Finkenwerder, a municipality in the west of Hamburg. These had two lift points with lifting capacities of 37 500 and 50 000 kg respectively. It was thus possible to lift even heavy flying boats onto trailers. A tractor then towed the seaplane to the nearby harbor basin, where a tower crane took over the task of placing the flying boat on the water. The existing crane lacked the required lifting power, however, and initially a tower crane had to be rented from the Port of Hamburg warehouse. Another technical advancement was the building of a modern slipway. An electrically-powered inclined lift lowered the flying boat and its transport wagon (slip wagon) into the water. This method of bringing a flying boat to the water had been used by a number of navies during the First World War. For longer periods on land the BV 222 was placed on a specially-designed docking trolley.

In addition to the production halls, the Finkenwerder facility also included a generously-proportioned design office with attached premises for the construction of mockups 1350 square meters in size and a small wind tunnel.

Development of the Blohm & Voss BV 222, which began in 1938, was a real challenge for the members of the Finkenwerder factory. With a wingspan of 46 meters, a length of 36.5 meters and a gross weight of between 36000 and 49000 kilograms, the BV 222 flying boat was among the largest and heaviest aircraft of its day. Nevertheless, the company, which was then a minor player in the German industry, was able to complete the construction program rapidly.

A few weeks after the start of construction work on the BV 222 V1, work also began on the V2 and V3. Construction of the first prototype began in Finkenwerder in August 1938 and proceeded apace until the outbreak of the Second World War on 1 September 1939.

In order to allow the BV 222 V1 to be used by the Luftwaffe as a transport flying boat, a 2.3 x 2.1 meter freight door was installed on the starboard side of the aircraft.

Blohm & Voss developed a new method for quantity production of the BV 222, the so-called "1 : 1 Outline". It consisted of full-size ink drawings of the components on transparent foil. Not only could the drawings, some of which were meters in size, be reproduced as needed in a copy machine, but they could also be placed directly on the metal to be worked on. Dr.-Ing. Richard Vogt thus pursued the development philosophy originated by him on the BV 222. This meant that the flying boat had a highly-loaded, narrow fuselage and was stabilized on the water by stabilizing floats, which in this case were slit and retractable. The wing was straight and employed the tubular spar system successfully used in the three-engined Ha 138, the four-engined Ha 139 and the twin-engined Ha 140. Within the tubular spar were six fuel tanks. Other features of the design were an at first glance conventional empennage and a propulsion system consisting of six 1,200-hp Bramo 323-R Fafnir nine-cylinder radial engines. It was originally planned to install 1,000-hp BMW 132H radial engines in the first prototype.

The configuration of the "boat hull" was determined through numerous towing trials carried out at the Seaplane Institute of the German Aviation Research Institute (DVL) in Hamburg. Aerodynamic investigations, however, were conducted in the DVL's wind tunnel in Göttingen. Controlling such a large and heavy flying boat posed particular problems. Here, too, however, Dr.-Ing. Richard Vogt found a solution. The rudder was controlled by two tabs (servotabs) in the trailing edge. The pilot controlled the tabs by means of pedals. Vogt designed multi-part elevators. The actual elevator was located inboard, with an auxiliary elevator in the middle and a mass-balanced landing flap outboard. The inner and central surfaces were fitted with tabs.

Its camouflage scheme clearly visible, the BV 222 V1 over water during a test flight.

The BV 222 V1 tended to roll while water taxiing. This XXXX was never eliminated, even in later versions of the flying boat.

The Blohm & Voss BV 222 V1 was given the new code X4+AH... 3vund ...

...and on 10 May 1942 was attached to Lufttransportstaffel-See 222.

All Blohm & Voss BV 222s operated by LTS 222 were given a special marking in the form of a code consisting of the letter "S" plus the number of the prototype on the vertical tail surfaces. The V1 in the photograph bears the code S 1.

The landing flaps were operated by electric motors. Thanks to these landing flaps, there was no need for a variable-incidence tailplane for trimming the flying boat.

A fixed slot extending over about half of the elevator's leading edge prevented flow separation. This control layout was new, but here too Dr.-Ing. Richard Vogt placed primary importance on safety. The third prototype of the Blohm & Voss Ha 140 floatplane (D-AMME) served as test-bed for the BV 222. The Blohm & Voss Ha 140 was the product of a specification issued by the State Ministry of Aviation in 1935 for a high seas multipurpose aircraft. In September 1937 the specification was changed to a twin-engined floatplane which could also be used as a torpedo-bomber or conventional bomber. After three prototypes had been built, further development was cancelled in favor of the Heinkel He 115.

A wing with a length of 2.25 meters and a chord of 1.9 meters with the new control surface layout was placed vertically on the aircraft's fuselage, so that the effectiveness of the control system could be tested under flight conditions.

While the first prototype was under construction, Blohm & Voss had a full scale wooden mockup of the BV 222 built in Finkenwerder. Among other things, it was used to investigate the interior fittings, which were documented in a series of photographs.

On 16 July 1940 members of Deutsche Lufthansa inspected the mockup and requested a number of changes. After these were completed, on 7 August 1940 the airline gave its approval to the modified interior layout of the flying boat. Implementation of the changes requested by Deutsche Lufthansa never happened, however, as by then it was clear to all involved that the BV 222 would never be delivered to Deutsche Lufthansa, for the Luftwaffe was in desperate need of all available transport space on account of the war.

Work on the BV 222 V1 progressed rapidly until the outbreak of war on 1 September 1939, when production of the Blohm & Voss BV 138 was given priority. Production of the BV 138 A-1 began at the end of 1939 and drew away workers, which had a negative effect on construction of the BV 222 prototype. This was the main reason why the Blohm & Voss BV 222 V1 was not completed until the end of August 1940, just under a year after the war started. The company's next priority was to begin flight testing as soon as possible. On 7 September 1940, after thorough preparations, Flugkapitän Helmut Wasa Rodig, Blohm & Voss' chief pilot, lifted the BV 222 V1, Werknummer 222/365 (D-ANTE) off the Elbe on its maiden flight. The aircraft's builders were less than satisfied by the results of the sixty-minute flight. It was unstable in level flight and also rolled while taxiing on the water. The following months until the end of the year were filled with more test flights and modifications, until in December a heavy buildup of ice on the Elbe brought test flying to a halt. The serious shortage of large-capacity transport aircraft forced the Luftwaffe to press the BV 222 V1 into military service as soon as possible, and a "long-range transport flight program"

A BV 222 of Lufttransportstaffel-See 222, probably the V1, in a Mediterranean harbor.

between Hamburg and Kirkenes in Norway was proposed. During this special program further testing of the BV 222 V1 went hand in hand with urgently needed transport flights for the Luftwaffe.

Calculations revealed that design of the Blohm & Voss BV 222 had required 270,000 design hours costing just under one-million Reichsmark. From today's point of view this was a minor sum. For the later BV 222 C series, in 1943 Blohm & Voss quoted the RLM a price of 1.2-million Reichsmark per aircraft. This price did not, however, take into account the propulsion system, which was provided by the customer, the aircraft's armament and other equipment. A completely equipped BV 222 C, on the other hand, was estimated to cost 1.7-million RM.

The Blohm & Voss BV 222 "Wiking" as a Transport Aircraft in Military Service

The Blohm & Voss BV 222 V1 flying boat, originally conceived for civilian service, first flew on 7 September 1940 and soon entered military service due to the exigencies of war under the direction of Flugkapitän Helmut Wasa Rodig, Blohm & Voss' company pilot. Contrary to requests by the Kriegsmarine to use the flying boat as a long-range reconnaissance aircraft, in the beginning it was employed exclusively for transport purposes. In the spring of 1941 a 2.3 x 2.1 meter freight door was installed on the starboard side of the flying boat and camouflage was applied to its upper and lower surfaces. The camouflage scheme used for German maritime aircraft consisted of three basic colors: RLM 72 (green), RLM 73 (green) and RLM 65 (pale blue). In contrast, Luftwaffe land-based aircraft were finished in a scheme of RLM 70 (black green), RLM 71 (dark green) and RLM 65.

In the end the aircraft's civil registration D-ANTE gave way to a new military Stammkennzeichen CC+EQ. Despite its military appearance, the aircraft initially remained unarmed. The V1 made its first flight in its new "camouflage dress" on 14 May 1941. At the controls for the seventy-minute flight was Flugkapitän Rodig once again. On 8 July of the same year, Rodig completed his "Guidelines for the Employment of the Blohm & Voss BV 222 V1 as a Transport". The document specified a takeoff weight of 43500 kg, including a 9400 kg payload, and a range of 3000 km for the aircraft. An increased payload of 10500 kg could be carried over a shorter distance. This formula also worked in reverse, and a reduced payload resulted in a considerable improvement in range. Helmut Wasa Rodig established a range of 6000 km for the BV 222 V1 with a takeoff weight of 45000 kg and a payload of "just" 4000 kg. The aircraft's maximum range of 6500 km, on the other hand, was based on complete use of the fuel capacity of 14750 kg and a maximum gross weight of 45500 kg. Rodig's report included typical loads for the V1, for example five Junkers Jumo 211 12-cylinder liquid-cooled engines (as used by the Junkers Ju 88) with their propellers, 70 fully-equipped troops, or loads with a total weight of up to 9 400 kg. According to Fliegerkorps Rodig's report, the BV 222 V1 required a minimum water depth of three meters and, fully loaded, a takeoff distance of 3500 meters and a landing distance of 1000 meters. Fuel consumption at a speed of 300 kph was 1100 liters per hour, while oil consumption during the same period was between 30 and 35 liters.

The Blohm & Voss BV 222 V1 carried out its first transport mission from 10 to 12 June 1941, flying from Seedorf on the Schaalsee, approx. 45 km east of Hamburg, via Friedrichshafen and the Lago di Bracciano near Rome to Augusta, a port in Sicily located about 30 km north of Syracuse. The five-man crew consisted of Flugkapitän Hans Wasa Rodig, his copilot Flugkapitän Hans-Werner von Engel and members Brackwitz, Dielewicz and Schneider. For this flight the aircraft carried special fuel pumps to ensure rapid refueling at its destination. On 10 June the V1 landed

The BV 222 V1 during loading trials in Hamburg-Finkenwerder. Cameras on special mountings in the nose area were installed to document flight trials.

28 October 1941: The Blohm & Voss BV 222 undergoes an engine change in Piraeus harbor, Greece.

Mechanics prepare to remove the Bramo 323 R Fafnir radial engine. A hoist has been positioned to raise the engine.

at Friedrichshafen on Lake Constance to pick up its cargo—large and small crates containing engine and transmission parts for half-tracked vehicles from the Maybach company located there. It was soon discovered that it was impossible to load several of the large, heavy transport crates from the waiting boats into the flying boat lying at anchor. The situation was further complicated by unusually rough waters on Lake Constance caused by a persistent storm. The loading work was finally halted on the evening of 10 June and during the night the contents of the too large and heavy crates were repacked in smaller, easier to handle crates. The next day the loading of the BV 222 V1 was completed successfully. Meanwhile two Junkers Ju 52/3m transports had delivered additional crates containing tank shock absorbers on behalf of the Air Transport Group (LVG). Five hours later loading of the flying boat was completed and the total weight of its load was 9200 kg. During the subsequent takeoff from Lake Constance, which took about one minute, the altered center of gravity caused porpoising which almost resulted in an aborted takeoff. Flugkapitän Rodig was finally able to lift the flying boat off the water and set course for Rome. The V1 crossed the Alps at an altitude of 3300 meters and at 15:37 it made an en route stop on Lago di Bracciano, located about 35 km northwest of Rome, where the Regia Aeronautica had a seaplane base. This stop was intended to avoid an overnight stop in Augusta, Sicily, as the port was a frequent target of enemy attacks. After landing on Lake Bracciano, a check revealed that one cylinder of engine No. 6 had been damaged. Repairs were carried out with equipment on the flying boat during the night of 11-12 June 1941. After a second night with little sleep for the crew, the next day the V1 completed its flight and landed safely in Augusta. The aircraft's cargo was unloaded, 7000 kg of fuel was taken on, and that same day the BV 222 V1 took off without payload to begin the 1880 km return flight to its home base on Lake Schaal. The flight took five hours and 24 minutes at an average speed of 348 kph.

During its first transport mission from 10 to 12 June 1941 the V1 flew 3960 km with a total flying time of 12 hours and 37 minutes, resulting in an average speed of 314 kph. Depending on the route segment, the Fafnir engines consumed between 920

The Blohm & Voss BV 222 V2 initially wore the code CC+ER.

In Piraeus harbor the V2 is brought ashore for overhaul.

LUFTWAFFE AIRCRAFT Profile

During a night landing in Piraeus, Athen's seaport, in February 1943, the V1 struck a sunken ship lying just beneath the surface and sank.

222 V1 still tended to roll while taxiing on the water as well as continued problems with the Bramo 323 R Fafnir engines. On 28 October 1941, while the BV 222 V1 was anchored to a buoy in Athens harbor, the number four engine had to be changed with the help of a crane and a floating pontoon, a process that proved rather difficult. The engine was replaced, however it was realized that such an engine change was rather risky on account of various problems associated with removing the propeller and handling the removed engine on the water. For this reason it was decided that future engine changes should be carried out on land whenever possible.

While the BV 222 V1's early transport flights to Norway had been flown mostly and 1230 liters of fuel. After the bad experience with the first loading of the V1 on Lake Constance on 10 June 1941, boats were no longer used to deliver cargo to the aircraft, instead the cargo was brought to the flying boat on a floating pontoon and then stowed on board with the help of two loading bridges and hoists. After this first transport mission in June 1941, it was planned that, after several short test flights between Lake Schaal, Travemünde and Finkenwerder in the following month, the BV 222 V1 would carry out a series of transport flights between Germany and Kirkenes in the north of Norway with Flugkapitän Hans-Werner von Engel at the controls.

The Blohm & Voss BV 222 V1 completed a total of seven supply flights to Kirkenes between the 3rd and 29th of August 1941. The aircraft covered about 30000 kilometers in 120 hours of flying time. The flying boat delivered 65000 kg of supplies and returned a total of 221 wounded to the homeland. The aggressive saline sea water and the barnacle buildup on the fuselage affected the aircraft so seriously that it had to undergo a period of maintenance. After its Norwegian operations the BV 222 V1 underwent a thorough overhaul. It was then sent to Athens, for the Africa Corps also needed large daily deliveries of supplies, not all of which could be transported by sea. In addition to numerous Junkers Ju 52/3m and several large-capacity Messerschmitt Me 323 Gigant transports, aircraft such as the Junkers Ju 90 and the Blohm & Voss BV 222 also had to be used for supply flights. These went hand in hand with further operational trials, which revealed that the BV

The BV 222 V4 in formation with another Wiking over the Mediterranean.

The BV 222 V5 also served as a transport aircraft with LTS 222 from July 1942 to May 1943.

12

The Blohm & Voss BV 222 V8...

...on its slipway somewhere in the Mediterranean.

over secure territory, in the Mediterranean the flying boat encountered completely different operational conditions. British long-range fighters prowled the skies, hunting the German transports and generally making their life difficult. Beginning on 16 October 1941 the BV 222 V1 began a shuttle service between Athens and Derna, Libya. Most of the time it was escorted by a pair of Messerschmitt Bf 110 fighters. At about the halfway point in the flight, the first pair of fighters turned back and another pair of Bf 110s from Africa escorted the flying boat to its destination. This procedure did not always work, however, and sometimes the escort fighters failed to show up at the rendezvous point. The unarmed Blohm & Voss BV 222 V1 was then forced to fly the rest of the way to its destination alone with no fighter escort. Luck was with the flying boat in this and the following year, however, and no enemy aircraft were encountered. From 16 October to 6 November the V1 flew the route between Athens and Libya a total of seventy times, delivering 30000 kg of supplies and evacuating 515 wounded. There are unconfirmed reports that a sort of gentleman's agreement with the British existed at that time, not to attack the BV 222 V1 on its return flights from Africa to Athens as its cargo of wounded also included British prisoners of war. Whether such an agreement existed or whether it was simply coincidence is unclear, however it is a fact that the BV 222 V1 was never attacked during its return flights from Africa in those months.

As a troop transport the BV 222 V1 could carry a total of 92 fully equipped tro-

Freight loading door open, a Wiking of LTS 222 waits for its cargo, this time consisting of members of the Wehrmacht.

The BV 222 V8 was the last example...

The Blohm & Voss BV 222 V9 later became the C-09 and the first C-series pre-production machine.

ops, while in the casualty evacuation role it could accommodate 72 litters.

The overall assessment of the BV 222 V1's initial operations was positive. Even if longitudinal stability and the flying boat's power plants were not entirely satisfactory, its performance was impressive. Maximum speed at an altitude of 4500 meters was 385 kph and its range of 7000 km made the flying boat indispensible for long-range operations.

In the winter of 1941-42 the BV 222 V1 was in Travemünde for general overhaul and it was intended that the flying boat should subsequently form the foundation of the new Lufttransportstaffel-See 222 (LTS 222), which was part of KGzbV 2.

The V1 was also fitted with defensive armament. This consisted of a 7.92-mm MG 81 in the fuselage nose, two rotating turrets on the fuselage spine each with one 13-mm MG 131 and four more MG 81s in positions in the fuselage sides. As well, the BV 222 V1's code was changed to X4+AH and LTS 222's code S 1 was painted on the tail section.

The Blohm & Voss BV 222 V2 was fitted with a strengthened hull bottom with five auxiliary steps directly behind the main step, in order to improve the flying boat's takeoff characteristics. With the growing requirement for transport aircraft, it was understandable that there would be pressure for the V2 (Werknummer 222/366) to quickly enter service. There was, however, disagreement between the navy and the air force as to how the BV 222 flying boats should be used.

While the Luftwaffe continued to view the type in the role of transport, the Kriegsmarine wanted to see these flying boats fitted with the necessary equipment and used as long-range reconnaissance aircraft over the North Atlantic. According to a memo by the Minister of Air Armaments dated 1 May 1941, the second prototype was supposed to be delivered to Kampfgeschwader 40 (KG 40). The BV 222 V2, which bore the code CC+ER on its maiden flight on 7 August

In this in-flight photo of the BV 222 V8 the white S 8 code is clearly visible on the vertical tail, identifying the flying boat as belonging to LTS 222.

During operations in the Mediterranean theater the BV 222 was in constant danger of being attacked by British long-range fighters.

For maintenance and repair work the Wiking flying boats were either brought ashore...

...or placed in a floating dock.

1941, was fitted with defensive armament consisting of a 7.92-mm Mauser MG 81 in the fuselage nose, two Rheinmetall-Borsig 13-mm MG 131 in ring mounts on top of the fuselage and four more MG 81s in positions in the fuselage sides. As well, an approximately 3.6-meter-long weapons gondola was mounted beneath each wing between the outer two engines. Each housed two remotely-controlled MG 131s, one firing forwards and one rearwards.

The reason for this unusual gondola-mounted armament was operational experience, which had shown that the Blohm & Voss BV 222 was unprotected against attacks from the rear, even at low level. One of its principal opponents in the Mediterranean, the Royal Air Force's Bristol Beaufighter, could fly closer to the surface of the water than the BV 222 on account of the latter's much greater height. The height of the BV 222 was 10.9 meters, whereas that of the Beaufighter was 4.82 meters on the ground and just 3.25 meters in level flight with the undercarriage retracted.

At such low heights the defensive positions in the fuselage sides and spine could not reach the Beaufighter, allowing the twin-engined fighter-bomber to conduct its attacks against its much larger opponent unhindered. Flight tests in Travemünde, however, revealed that the gondolas installed beneath the wings caused a marked deterioration in the BV 222's handling. The additional drag also reduced the aircraft's speed to the point that their use could not be justified. Both underwing weapons installations were therefore removed.

These modifications to the second prototype and its armament delayed its handover to the Luftwaffe, as did a fracture of the tubular spar, which halted acceptance flights for a long time. Ultimately the Luftwaffe won out over the Kriegsmarine, and on 10 August 1942 the BV 222 V2 (X4+BH) was delivered to Lufttransportstaffel-See 222 (LTS 222), which by then had been formed. Commanded by Hauptmann Fritz Führer a former Flugkapitän with Deutsche Lufthansa the Staffel was part of an air transport group that also included Lufttransportstaffel 290 (LTS 290), which was equipped with the six-engined Junkers Ju 290 land-based transport. The entire air transport group was under the command of Major der Reserve Rudolf Krause, also a former Lufthansa pilot. Lufttransportstaffel-See 222's first base of operations was Tarent in Italy.

In addition to the BV 222 V1, which had been fitted with defensive armament similar to that of the V2 and had been attached to Lufttransportstaffel-See 222 since 10 May 1942, LTS 222 also had available the BV 222 V3 (Werknummer 222/439, Stammkennzeichen DM+SD). The V3 had made its first flight on 28 November 1941 and on 9 December of the same year had been taken on strength by LTS 222 as X4+CH. According to many sources, LTS 222 considered the V3's armament to be unnecessary and removed all of the machine-guns except for the MG 81 in the nose. Between January and March 1942 the BV 222 V3 made a total of 21 supply flights from Tarent and Brindisi to Tripoli. In February 1943 the BV 222 V3 moved to Travemünde, where it was converted, fitted with defensive armament again, and subsequently attached to the Fliegerführer Atlantik in Bicarosse, France as a long-range reconnaissance flying boat.

Production of the BV 222 had by then reached a continuous state, although it could not be described as mass production. Lufttransportstaffel-See 222 received additional prototypes for operational purposes during the course of the year the BV 222 V4 (X4+DH) on 20 April, the BV 222 V5 (X4+EH) on 7 July, the BV 222 V6 (X4+FH) on 21 August and the V8 (X4+HH) on 26 October. The BV 222 V5 is known to have worn the LTS 222 marking S 5 on its fin.

As of 20 January 1942 the name "Wiking" proposed by Blohm & Voss was made official. The Blohm & Voss BV 222 V4 was the first aircraft to be fitted with a modified horizontal tail, which was supposed to be tested for the following Blohm & Voss BV

LUFTWAFFE AIRCRAFT Profile

A Blohm & Voss BV 222 A in front of its slipway. Photo: Peter P. K. Herrendorf

The Blohm & Voss BV 222 V7 on its docking trolley.

238. The aircraft bore an additional "X" marking on both sides of the forward fuselage.

The BV 222 V4 (X4+DH) made an especially dangerous flight on 10 December 1942. It was en route across the Mediterranean with the BV 222 V1 (X4+AH) and the BV 222 V8 (X4+HH) from Tarent to Tripoli, when the three flying boats encountered British long-range fighter-bombers. The three Bristol Beaufighter VIs of No. 227 Squadron, Royal Air Force had taken off from Malta on an escort mission at 07:35. In the course of their morning mission the British pilots spotted the three large German flying boats below them, heading south at a height of about five meters above the sea. In the attacks that followed, the British fighter-bombers went as low as about two meters above the water. As a result they found themselves in the blind spot of their much larger targets. From this position Flight Lieutenant Rae opened fire on the BV 222 V8, while his squadron mates attacked the other two Wikings. The four 20-mm cannon in the fuselage nose and the six .303-inch machine-guns in the wings of the Beaufighter had a devastating effect on their target.

The three German flying boats tried to give each other mutual protection, however the close-range fire from Rae's Beaufighter soon had its effect on the BV 222 V8, which

The Wiking's two bracing floats were positioned 15.5 meters outboard of the flying boat's centerline.

A formation of Blohm & Voss BV 222s over Finkenwerder.

As it flies overhead, this Wiking displays its impressive wingspan of 46 meters.

crashed into the sea and exploded. Several weeks earlier, in November 1942, the BV 222 V8 had sustained serious damage (40%) as a result of an engine fire and had only returned to service a few days earlier following repairs. Among those killed when the flying boat was shot down on the morning of that 10 December was the Knight's Cross's wearer Hptm. Wolf-Dietrich Peitsmeyer. He was a passenger on board the BV 222 V8 and was on the way to his unit, I. Gruppe of Stukageschwader 2, stationed in Tripoli.

In addition to shooting down the BV 222 V8, Flt.Lt. Rae also claimed a second BV 222 as a "probable". This second "probable" victory was the Blohm & Voss BV 222 V4, which was hit twice in the Beaufighter's attack. Though seriously damaged, the V4 managed to reach Tripoli and land safely. The BV 222 V1 was more fortunate than its fellows and escaped the attacks by the Beaufighters entirely. It arrived at its destination undamaged. It seemed that the three BV 222s were easy prey for the trio of No. 227 Squadron Beaufighters on this day, but perhaps not, for one of the British fighters, flown by Flight Sergeant Day and Flt.

Sgt. Featherstone, failed to return and another of the three Beaufighters was reported "damaged in aerial combat". According to British sources the three Bristol Beaufighter VIs of No. 227 Squadron were involved in another action after their attack on the three BV 222s. There opponent was a lone Fw 200 Condor. It was therefore possible that the Beaufighter crewed by Day and Featherstone was lost in the second action that day.

The Blohm & Voss BV 222 V4, which was damaged in the attack by the Beaufighters on 10 December 1942, was subsequently repaired and from May 1943 served as a reconnaissance flying boat under the Fliegerführer Atlantik.

After the loss of the BV 222 V8, Hauptmann Fritz Führer, the commander of Lufttransportstaffel-See 222, wrote: "... It has now become obvious that the BV 222's defensive armament is totally inadequate. Low-level flight, previously the best type of defense, no longer has any value. Any enemy fighter can position itself three meters lower than the flying boat and cannot be hit by the existing defensive weapons. Furthermore close formation flying by the BV 222 makes rapid evasive maneuvers almost impossible ..."

Aerial reconnaissance by the Royal Air Force had meanwhile given the British a clear picture of the routes and operational schedules used by LTS 222, making possible precise interception missions by their long-range fighters. German awareness of this and the costly flight by the three Wikings on 10 December 1942 ultimately resulted in the flying boats operating mainly by night.

Lufttransportstaffel-See 222 had previously lost the BV 222 V6 (X4+FH). At 13:00 on 24 November 1942, three Beaufighters of No. 272 Squadron, RAF were flying east along the coast of Tunisia. About 60 km from the island of Linosa, the British pilots sighted a lone Blohm & Voss BV 222, which was en route from Tarent to Tripoli, and attacked immediately. Flying Officer E.E. Coate attacked the flying boat from the rear and out of the sun. His first burst of cannon fire blasted pieces from the Wiking's fuselage and a second burst set the three engines on the left wing and the fuel tanks on fire. The flying boat lost height rapidly and struck the surface of the Mediterranean about halfway

Each of the BV 222 A's Bramo 323 R Fafnir engines produced 1,200 hp, giving the aircraft a maximum speed of 345 kph.

between Linosa and Pantelleria. It bounced into the air and flew for some distance. In the process the V6 lost its left wing, made a half roll, struck the surface of the water again and exploded. The aircraft's crew of eight and the fifty passengers on board were all killed. Shortly after this dramatic event three Messerschmitt Bf 109s arrived from a base in Africa and drove off the three Beaufighters of No. 272 Squadron.

To avoid the danger of additional losses, the leaders of KGzbV 2, to which LTS 222 was attached, ordered a change in air routes.

But enemy action did not account for all of the Wiking losses. In February 1943 the Blohm & Voss BV 222 V1 (X4+AH) made a water landing in Piraeus, the port of Athens, in total darkness on account of an air raid alert. The crew missed a warning buoy in the harbor and the aircraft collided with a sunken ship lying just under the water. The flying boat's hull was ripped open and it sank soon afterwards.

The situation of the German transport commander in the Mediterranean was now hopeless. On 5 April 1943 the Allies launched "Operation Flax", the interdiction of sea and air traffic delivering supplies to the German Africa Corps. This led to crippling losses on the German and Italian side, and in the following three weeks many German and Italian transport aircraft were lost. The two tragic climaxes of "Operation Flax" occurred on 18 and 22 April 1943. The first of these two big battles took place on Palm Sunday and was begun by four squadrons of American Curtiss P-40Fs. This attack on 65 Junkers Ju 52/3m transports, which went down in history as the "Palm Sunday Massacre", was the greatest success by the Curtiss P-40 during the Second World War and cost the Luftwaffe 59 of its trimotor transport aircraft and 14 escort fighters.

The second large air battle occurred just a few days later on 22 April 1943. That day about twenty Messerschmitt Me 323 Gigants of Transportgeschwader 5 ran into two squadrons of Spitfire Vs and IXs and soon afterwards three squadrons of Curtiss Kittyhawks, all of the South African Air Force. Sixteen or seventeen of the six-engined transports and ten German and Italian fighters were lost in this engagement. Tactically, TG 5 ceased to exist that day.

In addition, many Axis aircraft were more or less severely damaged and also put out of action. Much worse than the loss of equipment were the deaths of the flight crews. The Luftwaffe's transport service was never to recover from this latest bloodletting (after Crete and Stalingrad). On 13 May 1943 the German Africa Corps surrendered, ending hostilities in North Africa.

The remaining serviceable BV 222s in the Mediterranean had already taken on new tasks by then. They were now flying transport missions to the Crimea, in the area of the Kerch Peninsula, which separates the Black Sea from the Sea of Azov. The Kerch Peninsula, with its port city of the same name, had been taken by the German XXXXII Army Corps in November 1941. At the end of the following month the Soviets launched a landing operation and retook the peninsula, however it fell to the Germans again in May 1942. The BV 222 flying boats were ideally suited to supply the German military on the peninsula. These operations soon came to an end, however, for Groß-admiral Karl Dönitz, commander-in-chief of the German Navy since 31 January 1943, had convinced Hitler to order three BV 222s transferred to his command for use as reconnaissance aircraft.

Before their transfer to the Flieger-führer Atlantik, in March 1943 the BV 222 V5 also operated in Finland. The Wiking was given a temporary winter camouflage finish, becoming the only Wiking to be so painted. It is certain that the BV 222 V5 was stationed in the Finnish port of Petsamo in March 1943. It is uncertain, however, whether the flying boat was used solely as a transport aircraft or also participated in operations by the Kriegsmarine and Luftwaffe against the Allied convoy JW.53 as a reconnaissance aircraft.

Depending on load, the BV 222's draught was between 1.22 and 1.45 meters.

The Blohm & Voss BV 222 Wiking as a Long-Range Reconnaissance Aircraft over the Atlantic

The success of the German submarine campaign depended in large part on seamless aerial reconnaissance, however the German aviation industry proved incapable of producing a sufficient number of suitable aircraft. From the beginning of the Luftwaffe's creation, tactical considerations took pride of place, for armaments were conceived primarily for use against Germany's immediate neighbors. As a result the Luftwaffe lacked long-range aircraft which could take on the role of strategic reconnaissance aircraft. Development of the Heinkel He 177, a technically-complicated long-range bomber, was begun before the war started. Numerous problems resulted in a lengthy development process and many losses caused by technical problems or training. New developments such as the Messerschmitt Me 264 failed to reach quantity production on account of the worsening military situation or remained drawing board projects like the Focke-Wulf Ta 400.

As with its air transport service, the Luftwaffe was forced to turn to civilian developments for its long-range reconnaissance aircraft. The Focke-Wulf Fw 200 four-engined airliner proved best suited to this role. It is not surprising, therefore, that a fierce fight broke out between the Luftwaffe and Kriegsmarine over the few available long-range aircraft, for both arms urgently needed these aircraft. The Luftwaffe needed them as transports, bombers and, like the Kriegsmarine, reconnaissance aircraft. Without air reconnaissance the Kriegsmarine was almost powerless in the war against Allied convoys. As Reichsmarschall Hermann Göring's star sank as the war went on, Admiral Karl Dönitz, commander of German submarines since 1939, was able to steadily increase his influence on Hitler. He therefore gained a sympathetic ear when, in his new capacity as Commander-in-Chief of the Kriegsmarine, in February 1943 he requested the assignment of BV 222 flying boats to the Fliegerführer Atlantik for reconnaissance purposes. The Luftwaffe, which had previously blocked the transfer of these big flying boats to the Kriegsmarine, now showed itself surprisingly cooperative. Göring even became personally involved and sketched a picture of possible reconnaissance operations. In his opinion, in the event of bad weather or when no convoys were in sight, the BV 222 could "… set down off Greenland …" for one or two days before being able to intervene in events again. He overlooked two important points,

Although the BV 222 V3 appears to already be in service with LTS 222, it still wears its old code DM+SD.

This Blohm & Voss BV 222 C moves up its slipway with the aid of all six engines.

The primary role of the Blohm & Voss BV 222 C was that of a long-range reconnaissance aircraft for the Kriegsmarine, but – like its predecessors – it also performed transport missions.

LUFTWAFFE AIRCRAFT Profile

After the surprise attack on 1.(F)/Aufklärungs-Gruppe 129's base on the shore of Lake Bicarosse …

…by four DeHavilland Mosquito IIs of No. 264 Squadron, RAF on 20 June 1943, in which the Blohm & Voss BV 222 V3…

however. For one, the Blohm & Voss BV 222 had been designed as a civilian flying boat and was not designed for high seas operations, consequently it required a relatively calm Sea State 1 or 2 to take off and land. For another, the Bramo 323 R Fafnir engines had to be serviced after 25 hours of operations. This required a suitable ground organization and could not be carried out on the high sea.

In the view of the Luftwaffe, the introduction of the new BV 222 C series created expanded operational possibilities. The main difference between the C version and the preceding BV 222 V1 to V6 and V8according to many German sources these prototypes were designated BV 222 A was the propulsion system, which now consisted of Junkers Jumo 207 C diesel engines. One of the advantages of this diesel propulsion was that the diesel fuel housed in the unprotected tanks in the tubular spar was less flammable than the usual aviation gasoline, consequently the risk of fire was significantly reduced. An increase in range was also expected due to the lower fuel consumption of the diesel engines. Cruising performance at lower altitudes was also improved and it was also hoped that the BV 222 C would be able to refuel from German submarines on the high sea. In this case the aircraft's calculated range was as high as 2,588 nautical miles (4793 km). The resulting endurance of up to 33 hours at "economical cruising speed" made it entirely

…and V5 plus a Blohm & Voss BV 138 were sunk at anchor…

…the Commander-in-Chief of the Kriegsmarine, Grossadmiral Karl Dönitz, ordered branch channels built on the shore of the French lake to better protect the flying boat.

Bow and stern views of the Blohm & Voss BV 222 model in a typical smooth-water situation.

Front and rear views of the Blohm & Voss BV 222 model in a typical calm water situation.

possible to extend the time before overhaul to fifty hours for the diesel engines.

Relevant tests were carried out with the BV 222 V7, the prototype for the C series. These will be described in detail later.

After the two prototypes Blohm & Voss BV 222 V3 Werknummer 222/439 (X4+CH) and V5 Werknummer 222/000 0005 (X4+EH) had been converted for their new role in Travemünde, on 17 May and 12 June 1943 they were delivered to 1.(F)/Aufklärungsgruppe 129 stationed in Bicarosse on the French Atlantic coast.

The city of Bicarosse is located on the shore of the lake of the same name, approximately 70 km southeast of Bordeaux. The small Lake Bicarosse was an ideal base for the flying boats and was separated from the Bay of Biscay by a narrow land bridge just 8 km wide. Oberleutnant Möhring's logbook confirms a mission by the BV 222 V3 on 9 June 1943. At 14:00 he took off from Bicarosse on a 15-hour long-range reconnaissance sortie. The Royal Air Force soon learned about the transfer of the two flying boats to the Biscay coast, probably through their signals intelligence, and on 20 June 1943 four DeHavilland Mosquitoes of No. 264

Blohm & Voss BV 222 V8 with X4+HH registration, of LTS-Sea 222, with the 1942 paint scheme.

LUFTWAFFE AIRCRAFT Profile

The antenna state is a real masterpiece.

The upper weapon mount in detail.

The side weapon mount on for forward left side is also finely detailed. sehr fein detailliert.

The exits of the exhaust pipes were so painted as to look like rust.

The Model of the BV 222 in Detail

Rear view.

The radio and navigation antennas of the Bv 222.

LUFTWAFFE AIRCRAFT Profile

Blohm & Voss BV 222 C-1 with Diesel engines, TB+QM registration, equipment state circa 1943.

Blohm & Voss BV 222 B2 as X4+BH, with added winter camouflage for Operation "Treasure Digger" in 1944.

LUFTWAFFE AIRCRAFT Profile

The portrayal of an open motor gives life to a model. Naturally the detailing is extremely precise/ Note the signs of rust on the exhaust manifold.

Detail view of the right engines with effective raised engine hoods.

In this picture of the left support float, the hydraulic cylinder is especially easy to see.

The elaborate design of the divided support floats is nicely portrayed here.

The 1:72 scale model invites an effective size comparison between the BV 222 and a Jn 53 3m/Sea.

The Blohm & Voss BV 222's bracing floats were divided and with the help of electrically-powered cables retracted into recesses in the underside of the wing.

When the flying boat was in a level position the bracing floats were one meter above the surface of the water.

Squadron made a surprise attack on 1.(F)/AufklGr. 129's base. The attack destroyed both Wikings and a Blohm & Voss BV 138 that was lying at anchor on Lake Bicarosse. The destruction of the two flying boats was a painful loss for the German navy. Großadmiral Dönitz said that the destruction of the two Wikings "…is comparable to the loss of two cruisers in a naval battle…"

In the 1990s French divers of the Centre d'Essais (CEL) began recovering parts of the Blohm & Voss BV 222 V3 from Lake Bicarosse.

A curious footnote to the story of the BV 222 was that at the beginning of June 1943, despite the grave shortage of long-range reconnaissance aircraft, the Commander-in-Chief of the Luftwaffe issued an order for a BV 222 to be used exclusively for search and rescue operations. This order must come as surprising, for the aircraft's limited seaworthiness, which restricted it to landings in relatively calm seas, made it quite unsuitable for this task. The use of a BV 222 as an air-sea rescue aircraft was also questionable because at the same time there were four French-made Breguet 521 Bizerte flying boats in service with the Luftwaffe's 1. Seenotstaffel in Brest-Hourtin and these could operate even in Sea States 4 to 5.

After the successful attack on 1.(F)/Aufkl.Gr. 129 in Bicarosse by the Mosquitoes of No. 264 Squadron, the only surviving prototypes of the BV 222 were the V2 and V4, which were also earmarked for service as long-range reconnaissance aircraft. After the bitter experience of 20 June 1943, on the initiative of Großadmiral Dönitz, branch channels were laid out on the shore of Lake Bicarosse to provide better protection of the flying boats. The two surviving Wikings could be towed into the channels for maintenance and concealment.

Meanwhile the German admiralty (SKL) was planning the formation of a "Geschwader for extreme long-range reconnaissance", with the wording "extreme" apparently intended to convey an augmentation of the term "long-range". Six Blohm & Voss BV 222 Cs were supposed to be combined in a long-range reconnaissance Staffel by the end of 1943. It was also intended that the Staffel should operate land-based types such as the Junkers Ju 88 H, Heinkel He 177 and Junkers Ju 290 A, plus Blohm & Voss BV 138 and Arado Ar 196 seaplanes. The latter single-engined floatplane would be used for tactical maritime reconnaissance.

More aircraft were wanted, but building them was impossible because of the production situation that existed at the time. Heavy bombing raids on production sites, shortages of skilled workers and the absence of certain raw materials limited the output of large aircraft. Production of the Blohm & Voss BV 222 was also affected. The BV 222 C program was cancelled almost as soon as it began. On 3 August 1943 the decision was made at a meeting held by the Minister of Air Armaments to adopt the Emergency Fighter Program, which called for increased production of fighter aircraft in favor of other types. Construction of the Blohm & Voss BV 222 and the older BV 138 would be allowed to run down, while the BV 222's successor, the even larger BV 238, would not be put into production at all. According to the delivery plan of November 1941, a total of 86 BV 222s should have been built, with the monthly completion of up to four aircraft. Because of demands by the German admiralty, Luftwaffe Construction Program 223 of 1943 also included a small series of BV 222 Wikings for the Kriegsmarine. Five BV 222 Cs were supposed to be built by December 1943, 17 more in the following year and finally another 18 flying boats in 1945. Despite the BV 222 production halt in August 1943, work on the aircraft supposed to be completed by December 1943 nevertheless went on with a low priority until May 1944. As a result, only a few more Blohm & Voss BV 222 Wikings were completed, instead of the 86 aircraft originally planned.

The growing strength of the Allied defenses in the Atlantic meant that operational conditions there were no longer acceptable for the Blohm & Voss flying boats. This resulted in the decision to concentrate the BV 222 and BV 138 in Norway under Luftflotte 5. It was expected that these slow flying boats, which were now seen as having been rendered obsolescent by the latest developments, would find more favorable operating conditions there on account of the geographical conditions and the weaker enemy defenses. For this reason, effective 1 November 1943 with the arrival of the newly-assigned Fernaufklärungsgruppe 5 (FAGr. 5), the Fliegerführer Atlantik (commander of air forces Atlantic) was supposed to release all his Blohm & Voss BV 222s to Luftflotte 5.

The admiralty opposed this move and, at its urging, on 14 October 1943 the Luftwaffe Operations Staff decided that the BV 222s would remain in Bicarosse, even after FAGr. 5 was established there, and would be available to the Fliegerführer Atlantik.

A shortage of laborers prevented the ground organization in Bicarosse from being expanded, and consequently 1.(F)/Aufklärungsgruppe 129 even after the arrival of new C-series aircraft rarely had more than two aircraft serviceable at any one time.

Because of the war, the civilian B version of the BV 222 remained a drawing board project and the prototype for the BV 222 C series was the V7 (Werknummer 222/031 0007) with the code TB+QL. This aircraft was first flown on 1 April 1943 and on 14 April began its flight test program. As prototype for the C series, the BV 222 V7 was also designated BV 222 C-07. The Junkers Jumo 208 diesel engines producing 1,500 hp (1103 kW) envisaged for this version were not yet available, however, and the significantly less powerful Jumo 207 C (1,000 hp, 735 kW) had to be used in their place. It was soon discovered, however, that the Jumo 207 C was a sensitive power plant which was prone

All C-series Wikings were equipped with a folding nose for the loading of cargo. Seen here is the BV 222 C-09.

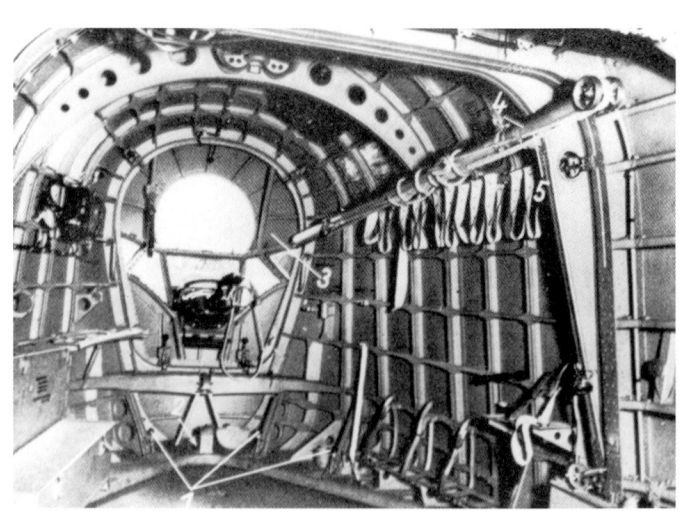

Forward view from the nose compartment, showing the folding nose in the closed position. The nose compartment extended from Frame No. 1 to No. 7.

to breakdowns. This problem was to accompany all C series BV 222s equipped with this engine for the entire time they were in service. The Wiking could also operate with takeoff-assist rockets. Given the cover designation R-Geräte (smoke devices), these rockets allowed the aircraft's gross weight to be increased to 51000 kg and in exceptional cases to 53000 kg. The designers had conceived a folding fuselage nose for the BV 222 B. Even though the BV 222 C was not being considered for use as a transport, this feature of the civilian BV 222 B was incorporated into its design. Other detail changes were made to the fuselage nose, resulting in the BV 222 C having a fuselage that was 50 cm longer, resulting in a length of 37 meters. Because of the military situation, testing of the BV 222 V7 was carried out over the Baltic Sea. Experiments were also carried out there with refueling the flying boat using diesel fuel from submarines.

When Holland was occupied, three Dutch O 21 class submarines fell into German hands. The Dutch Navy submarines were designated with a letter "O" followed by an Arabic numeral, for example O.21, if they were to be used in home waters. The Dutch submarines operated in the Dutch colonies, the so-called "colonial class", were designated with a "K" and a Roman numeral, for example K.XII. In 1940-41 the former Dutch submarines O.25, O.26 and O.27 were fitted out for a crew of 44 by the Kriegsmarine and returned to service as the UD.3, UD.4 and UD.5. Each was just under 78 meters long and weighed 8,881/1,380 GRT. In Dutch service the submarines had had a crew of sixty.

The submarine UD.4 was commissioned by Korvettenkapitän Brümmel-Patzig on 28 January 1941 and entered service with the Kriegsmarine, where it was used as a training and experimental vessel. From January 1943 to November 1944 it was attached to the 27th Submarine Flotilla in Gotenhafen.

The growing strength of the Allied defenses forced the Kriegsmarine to adopt new tactics. These included the refueling (called "oiling" by the navy) of U-boats by other U-boats. After conversion work on its upper deck, the UD.4 began refueling experiments in the late summer of 1943 which proceeded as follows:

"UD.4 sailed at around 2 to 3 knots (3.5 – 5.5 kph) on the surface ahead of the receiving submarine. Transfer of the oiling hose and a telephone cable was accomplished by means of a hawser and buoy. The receiving U-boat first picked up the buoy and then secured the hawser to the hull before connecting the telephone cable and the fuel hose. Both submarines subsequently dove to

This photo was taken from the Blohm & Voss BV 222 V7's cockpit, looking aft. Note the bottom of the HD 151/D power-operated dorsal turret.

30-35 meters, with UD.4 towing the receiving U-boat at about 3 to 4 knots. Because of the lack of a pumping system, the diesel fuel had to be forced through the hose with the aid of presswater. It required just under four hours to transfer 80 cubic meters of diesel fuel. After the maneuver was over both submarines surfaced and UD.4 retrieved the buoy, oiling hose, telephone cable and hawser.

In the course of training, UD.4 completed a total of about 220 transfers by the end of 1944. This submarine was also used in refueling experiments with the Blohm & Voss BV 222 V7, above water of course, but

LUFTWAFFE AIRCRAFT Profile

Another photo of the BV 222 V7. On the left is the "navigation cupboard".

The cockpit of the Blohm & Voss BV 222 V7.

On the left is the radio equipment and right the engine control panel of the Blohm & Voss BV 222 V2.

The cockpit instrument panel of the Blohm & Voss BV 222 alone had about 40 instruments and switches.

which were analogous to the transfer of diesel fuel from submarine to submarine. Parallel to this test the Kriegsmarine investigated the carriage of aviation fuel by submarine with the following results:

- Type VIID submarine
  23 tons in mine shafts

- Type IXC
  23 tons in ballast tanks

- Type XB submarine
  95 tons in mine shafts

The range of the Blohm & Voss BV 222 was also calculated, resulting in the following values for the BV 222 A:

- in Sea State 1:
  Fuel load   20000 liters
  Range   5500 km
  Penetration depth 2200 km

- in Sea State 2-3:
  Fuel load   8000 liters
  Range   2200 km
  Penetration depth 900 km

Thanks to the more economical diesel engines, better performance was expected from the Blohm & Voss BV 222 C:

- in Sea State 1 with use of takeoff assist rockets:
  Fuel load   17000 liters
  Range   6300 km
  Penetration depth 2500 km

- in Sea State 1 without takeoff assist rockets:
  Fuel load   13000 liters
  Range   3500 km
  Penetration depth 1400 km

- in Sea State 2-3 with a fuel load, performance was similar to that of the BV 222 A.

Experiments conducted with the BV 222 V7 were unsatisfactory, however. Water got into the fuel tanks resulting in the loss of three engines, and the Wiking had to be towed back to Gotenhafen. A similar refueling exercise in the Bay of Biscay was called off because of V7's bad experience. After operational trials in the waters of Siberia with Blohm & Voss BV 138 flying boats and U 255, a Type VII C submarine commanded by Kapitänleutnant Reche, yielded similar negative results, the idea of refueling flying boats from submarines on the high sea was finally abandoned. The notion of longer Atlantic operations with the Wiking was also unrealistic on account of its limited seaworthiness (Sea States 1 to 2).

Prior to this the notion of a possible attack on New York had been considered. In 1942 Adolf Hitler himself approached the General der Flieger, Generaloberst Hans Jeschonnek, with the idea. He was of the opinion that Blohm & Voss BV 222 flying boats could be refueled from submarines, making an attack on the American metropolis a possibility. The next month the possibility of attacking New York with one or two BV 222s was examined. The two flying boats would be refueled by two Type IX C submarines about 1000 kilometers off the American coast, after which they would bomb the Jewish quarter of the harbor area during the night. A third U-boat was to be stationed north of the Azores as an additional safety measure for the outward and return flights by the flying boats. Grossadmiral Dönitz believed that the operation was possible, but he expressed reservations over the diversion of the two IX C submarines, suggesting that they were capable of sinking between 60,000 and 80,000 gross registered tons of shipping in the same period. The staff of Minister of Air Armaments Erhard Milch was also convinced of the feasibility of the operation. On 28 July 1942, however, Generaloberst Jeschonnek rejected the idea as not feasible and, despite Milch's attempt to have Göring intervene, the idea of such an attack petered out. As well, it has never been explained how the BV 222 C, which never carried a bomb load even during its operations against Allied convoys in the Atlantic, was supposed to attack its targets in New York.

On 16 August 1943 the Blohm & Voss BV 222 V7 was assigned to the Fliegerführer Atlantik. For the German submarines and their long-range reconnaissance aircraft, the situation in the Atlantic took a drastic turn for the worse in those months because of the growing Allied material and technical superiority. Modern Allied aircraft equipped with radar, a field in which Great Britain was the world leader in the mid-1940s, and effective anti-submarine weapons made life difficult for the German submarines. During the war, electronic technology expanded beyond warning and fire-control systems into other areas. One of these was Asdic (Anti-Submarine Detection Investigation Committee), an active acoustic submarine detection device which posed a serious threat to German submarines and resulted in the destruction of many.

The utility room of a Wiking – here looking forward – extended from Frame No. 22…

…to Frame No. 26 (here photographed looking aft.

The utility room was also the domain of the "chief cook".

The crew rest room was located in the lower deck – here seen looking forward.

The same rest room of a Blohm & Voss BV 222 A photographed looking towards the rear.

The tail compartment of the flying boat began at Frame No. 44 and extended to the "hull tail".

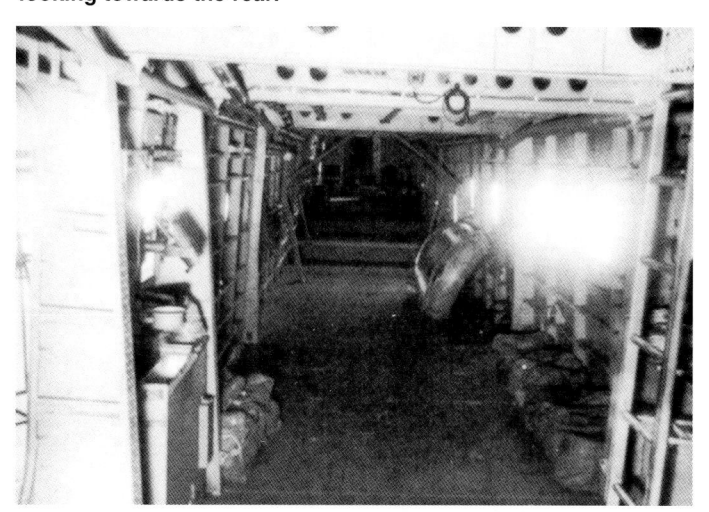

The spacious freight compartment of a Wiking, here photographed from the front…

The Kriegsmarine's long-range reconnaissance aircraft, on the other hand, were finding that a growing numbers of convoys were accompanied either by escort carriers or CAM ships equipped with fighter aircraft to defend against German aircraft. The British Catapult Armed Merchant Ships were merchant vessels which formed part of a convoy and had a Hawker Sea Hurricane mounted on a catapult for a one-time interception mission. If a German reconnaissance aircraft approached the convoy, the Hurricane was launched from its catapult and tried to shoot down or drive off the intruder before it could guide U-boats to the convoy. After its mission was complete or the aircraft ran out of fuel, its pilot had to ditch near the convoy so that he could be picked up.

As previously mentioned, this situation resulted in the idea of stationing the remaining Blohm & Voss BV 222s–the V2, V4, V7 and C-09–in Norway. There were two things wrong with this idea, however. For one there was no suitable base from which the flying

…and here from the rear.

The entrance to the forward...

...and to the rear crawlways of a Blohm & Voss BV 222.

The baggage compartment of a Wiking, photographed looking aft.

boats could operate and be supplied, and for another the German admiralty successfully vetoed the idea. And so it was that on 23 July 1943 the BV 222 C-09 (ex V9) was placed under the command of the Fliegerführer Atlantik. The V2 and V4 prototypes, which had been assigned to the Fliegerführer Atlantik in May 1943, underwent a thorough inspection and were delivered to 1.(F)/Aufkl.Gr. 129 in Bicarosse on 16 September 1943. A few days later, on 25 September, the BV 222 V2 (X4+BH) completed its first reconnaissance flight, which lasted 19 hours. By 8 October the two flying boats had flown seven missions, during which they located three convoys (MKS 25 on 29 September west of Lisbon, UA 2 on 3 October south of Ireland, and SC 143 on 8 October in the eastern Atlantic). The Wikings failed, however, to guide German submarines to these Allied convoys. In one case an unsuccessful attempt was made to direct the U-boat group "Roßbach" to one of the sighted convoys. These failures were homemade, as the example of Oblt. Möhring's reconnaissance flight in a BV 222 on 8 October shows. At 13:39 on that day he sighted the Allied convoy SC 143, consisting of 39 freighters, an escort carrier and nine escort vessels, at position 56º N 25º W and immediately transmitted its position by radio to the Fliegerführer Atlantik. There the report was passed to Kapitän zur See Godt for further action. Godt, however, did not believe the navigational accuracy of the report. He suspected that the convoy was approximately 185 km farther south and ordered his submarines there. As the U-boats' search for convoy SC 143 was too far south, they of course failed to find it and on 9 October the unsuccessful action was called off. It later turned out that the convoy was about 41 km further north than Möhring's report had stated. Nevertheless, the accuracy of the position information sent by the flying boat was an outstanding accomplishment and the responsibility for its failure to result in an operational success lay elsewhere.

Several days earlier four Fw 200 Cs had begun tailing convoy MKS.28 sailing north from Gibraltar but then lost contact. On 30 October a BV 222 of 1.(F)/Aufkl.Gr. 129 found the convoy again and began directing German submarines to its location. The convoy, made up of 60 ships and Escort Group B.1, was attacked repeatedly in the days that followed, resulting in losses on both sides. After this successful operation the Fliegerführer Atlantik made the following assessment:

"In repeated difficult missions on this aircraft type, the Staffel demonstrated good tactical training and dedication to duty. Keep up the good work!"

At noon on 26 November 1943 German long-range reconnaissance aircraft operating over the North Atlantic sighted convoy MKS 31, which was also accompanied by Escort Group B.1. On the following day a Blohm & Voss BV 222 of 1.(F)/Aufkl.Gr. 129 was able to remain in contact with the convoy which had been joined by convoy SL 140 and now consisted of 68 ships plus escorts of the 2nd and 4th Support Groups for five hours, guiding six German submarines to the area with homing signals. In the attacks that followed, the U-boats achieved little success, mainly due to the modern and efficient defense mounted by the Allied ships and aircraft.

All of the BV 222's missions over the Atlantic were flown from Bicarosse. The Fliegerführer Atlantik also wanted to operate shuttle missions between Bicarosse and Norway, however a search for a suitable landing site revealed that there were no seaplane facilities in Norway that were suitable for night or bad weather landings by the big flying boats. And so the shuttle missions had to be abandoned.

In addition to their defensive armament and FuG 216 Neptun rear-warning radar, the V2 and V4 had also been equipped with an extensive suite of electronic equipment and a FuG Hohentwiel maritime search radar, whose small antennas were grouped in the aircraft's nose. The FuG 200, which was also used by the Focke-Wulf Fw 200 C-4 and C-8 Condor and the Junkers Ju 88 H, Ju 188 and Ju 290 A, was capable of detecting ships at a range of 200 km. The defensive armament of the BV 222 C series consisted of combinations of gun positions in the nose, fuselage spine, fuselage sides and turrets on the wings accommodating weapons of various calibers MG 131 (13 mm), MG 151 (15 mm) and MG 151/20 (20 mm). Although this defensive armament seemed quite imposing, various circles within the air force and the navy regarded it as too light. Because of the military

A ladder connected the upper and lower decks of the Blohm & Voss BV 222.

A view of the lubricants control panel.

Here is the spar room, photographed looking forward.

The nose compartment of the Blohm & Voss BV 222 C-09.

circumstances, testing of the BV 222 C series was done in Germany's own, relatively safe "back yard" over the Baltic.

In October 1943 during a joint reconnaissance mission over the Bay of Biscay, the Blohm & Voss BV 222 V2 and V4 encountered an Avro Lancaster bomber. In the ensuing combat the British bomber was shot down by one of the flying boats. There is little to be found in British records about this Lancaster and the purpose of its flight. There is speculation as to whether the aircraft was on a ferry flight, a mine-laying mission or was possibly an Avro Lancastrian, a transport version of the Lancaster bomber.

By the end of 1943 the number of aircraft available to the Fliegerführer Atlantik was no longer sufficient to provide adequate marine reconnaissance and the sighting of an Allied convoy was largely a matter of chance. As well, the long-range reconnaissance aircraft had to be withdrawn from service at regular intervals for general overhauls or temporarily taken out of action for minor repairs. As a result of these interruptions in operations, the numbers of serviceable reconnaissance aircraft and the total number of aircraft available to the Fliegerführer Atlantik dropped sharply. On 11 November 1943, for example, the Fliegerführer Atlantik had just two Blohm & Voss BV 222 Wikings, of which one was serviceable. And of the Wiking's six-man crew, only five airmen were fit for duty at that time. Because of such, mainly technical, breakdowns of the overtaxed aircraft, which of course also affected the land-based aircraft of FAGr 5, many Allied convoys reached their destinations without being sighted and attacked. In the first two months of 1944, a total of 24 operational flights had to be aborted because of unexpected mechanical problems. As well as the Focke-Wulf Fw 200 C, Junkers Ju 88 H and Ju 290 A, the BV 222s based in Bicarosse were twice affected by such technical defects. By that time it is likely that only the Blohm & Voss BV 222 V2, V4, V7, C-09, C-010 and C-011 still existed, with the latter two Wikings stationed on the Baltic Sea at the end of 1943. The last two C-series Wikings, the C-012 and C-013, did not make their maiden flights until 23 November 1943 and 18 April 1944, consequently neither was able to have much of an effect on operations. Probably out of necessity, the Blohm & Voss BV 222 C-013 was powered by obsolete Junkers Jumo 205 C diesel engines (700 hp, 515 kW) instead of more powerful Jumo 205 D power plants. The latter were not available and so Blohm & Voss turned to the old Jumo 205 C engines. The Junkers Jumo 205 diesel motor was the successor to the Jumo 204, which had its origins in the 1930s, and was the forerunner of the Jumo 207 C engine used in the other BV 222 C flying boats. The Jumo 205 C had been used to power the Ju 86 D-1 in the 1930s and by the time it was installed in the BV 222 C-013 it was technically obsolete.

LUFTWAFFE AIRCRAFT Profile

MG 131 machine-guns were positioned in the fuselage sides of the Wiking flying boat…

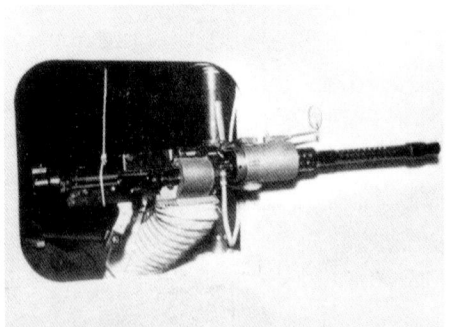

…in SL 131 mounts.

The side-mounted MG 131 machine-guns could be folded inwards for stowage.

The dorsal turret (B-1 Stand) on the forward fuselage spine – here on the BV 222 V4 – was armed with a MG 151/EZ cannon.

If required, a second power-operated turret could be mounted on the aft section of the fuselage spine (B-2 Stand). In this photo of the Blohm & Voss BV 222 V5 in service with LTS 222, however, an observation cupola has been fitted instead of the gun turret.

Additional defensive armament – here on the BV 222 V4 – included two wing-mounted turrets with MG 151/EZ cannon.

View of the rotating dorsal turret of the BV 222. The opening behind it was used for astral navigation during night flights.

34

The Allies' crushing superiority and the associated danger of fighter interception understandably reduced the number of missions flown by the Wiking flying boats. Another aircraft was lost on 8 February 1944, when the Blohm & Voss BV 222 V-010 was shot down near Bicarosse by a DeHavilland Mosquito II of No. 157 Squadron RAF flown by Flt.Lt. H.E. Tappin. No. 157 Squadron had received its Mosquito IIs in January 1942, becoming the first squadron of the Royal Air Force to operate the type.

On 22 March 1944 the Commander-in-Chief of the Kriegsmarine called off all operations against Allied convoys planned for the new moon period at the end of March. In addition to twelve U-boats, the Fliegerführer Atlantik had just a handful of Junkers Ju 88 Hs, eight Ju 290 As and one Blohm & Voss BV 222 available for reconnaissance duties. This small number of long-range reconnaissance aircraft made operations pointless. The last reconnaissance flight over the Atlantic by a Blohm & Voss BV 222 took place on 27 May 1944.

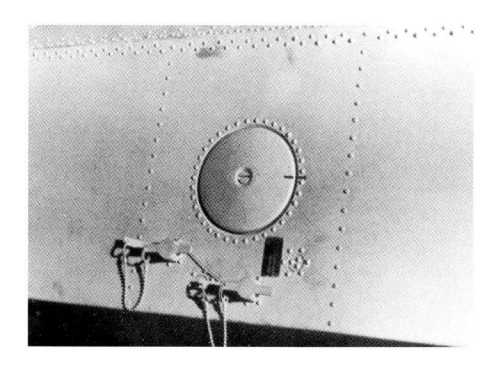

Connections for the flying boat's bilge system.

After the Allied invasion of France on 6 June 1944 (D-Day), operations against the Atlantic convoys from bases on the French coast were drastically curtailed, and soon afterwards the remaining four BV 222s stationed in Bicarosse and the III. Gruppe of KG 40 with its land-based aircraft were moved to Norway. There the Wikings were attached to Seeaufklärungs-Gruppe 130 and stationed in the ports of Tromsö and Sörreisa.

At the war's end British troops discovered the Blohm & Voss BV 222 V2, C-011, C-012 and C-013 undamaged in these two Norwegian ports.

Flugkapitän Möhring, who had since been promoted to Hauptmann and who had flown a total of 255 missions on the Blohm & Voss BV 222, was probably the foremost exponent of this flying boat. He flew his last operational flight, to Franz-Josef Land, shortly before the end of the war. The crew of Weather Station "Schatzgräber" (Treasures) located there had become ill with trichinosis. The men were originally supposed to have been flown out in a Fw 200 Condor, however one of the aircraft's wheels was wrecked in a landing near the weather station. Flg.Kpt. Möhring subsequently took off from Travemünde in the BV 222 V2 and dropped a replacement wheel for the Fw 200. The undercarriage was repaired and the Condor was able to complete its mission of recovering the members of the weather station who had fallen ill.

Weather information was of vital importance to operations by the Kriegsmarine and the Luftwaffe. For this reason Germany deployed weather ships and installed automatic weather stations on Spitzbergen and Bear Island. The establishment of manned stations was also required, however. One such operation was dubbed "Bassgeiger" (Bass Fiddle). On 14 August 1943 the German steamer Coburg, with 27 passengers on board, left Rostock, initially bound for Narvik. The ship's ultimate destination, however, was the east coast of Greenland, where the manned weather station "Bassgeiger" was supposed to be established. Just short of its destination, however, the Coburg became trapped in the ice. The vessel ultimately broke free temporarily, but in the end it had to surrender to the forces of nature and tie up on the permanent ice cover of Greenland. The crew

The first Wiking series was equipped with Bramo 323 R Fafnir radial engines. This photo depicts the Blohm & Voss BV 222 V1.

The C version of the Wiking was powered by Junkers Jumo 207 C diesel engines.

The Junkers Jumo 207 C power plant developed 1,000 hp for takeoff.

A Junkers Jumo 207 C engine of a BV 222 C with open access panels.

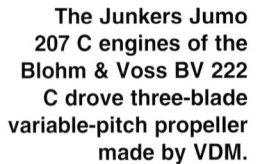

The Junkers Jumo 207 C engines of the Blohm & Voss BV 222 C drove three-blade variable-pitch propeller made by VDM.

of the Coburg subsequently erected two camps on the ice about 200 to 300 meters from their ship. Plans were made to fly in supplies and the E-Stelle See in Travemünde was tasked with making preparations. On 27 October 1943 pilot Adolf Mlodoch and his crew of 13 took off from Travemünde for Trondheim in the Blohm & Voss BV 222 C-09 and the following day flew on to Tromsö. In Tromsö the crew then had a rest break of about two weeks. Not until 17 November 1943 did the actual long flight to the members of weather station "Bassgeiger" in Greenland actually begin. The Blohm & Voss BV 222 C-09 (Werknummer 222/031 0009, TB+QM) was fitted with considerable defensive armament as well as the latest available navigational and electronic aids. These included the FuG 200 Hohentwiel search radar, for example. The supply flight to Greenland was also used to test this equipment under operational conditions. When it reached its destination, the flying boat was supposed to drop its supply canisters by parachute using the fuselage loading doors. First, however, the aircraft had to complete a long and dangerous flight during which it faced the constant threat of interception by Allied long-range fighters as well as technical and weather-related problems. Some time after the aircraft took off for Greenland the C-09's Number 1 engine failed. The reason was a fractured coolant line in the Jumo 207 C. A mechanic made his way down the crawlway in the wing and effected repairs in flight. No sooner was the Number 1 engine running again when a heavy ice buildup began to affect the entire aircraft. Pilot Mlodoch had no choice but to alter his route. Despite this weather-related detour, the aircraft finally reached its destination. But fate had yet another unpleasant surprise in store for the BV 222 C-09 and the members of "Operation Bassgeiger". An extreme bad weather front over the target area reduced visibility to almost zero, making it impossible to drop the load of supplies. Because of the poor visibility and drifting ice, the flying boat also could not land near the permanent ice. As well, the Wiking still faced a long return flight and did not have sufficient fuel to circle and wait for an improvement in the weather. For these reasons, the BV 222 C-09 and its frustrated crew had no choice but to set course for home. After a return flight lasting more than twelve hours, and having covered about 3000 kilometers, the flying boat arrived in Tromsö without further incident. It must have been small consolation to the crew of the C-09.

During the long weeks of their involuntary stay on the Greenland ice cap, the members of "Operation Bassgeiger" not only had to fight for survival against the forces of nature in that hostile environment, they also had to battle Allied forces. The enemy had become aware of the German activities on the east coast of Greenland and British troops travelling by dogsled launched a surprise attack on the two German camps on 22 April 1944. Leutnant Zacher, military commander of the German force, was killed, however the defenders successfully repulsed the British attack. It did, however, demonstrate the danger facing the members of "Operation Bassgeiger" and they were of course extremely happy when a Junkers Ju 290 was able to land and fly them home.

The Histories and Markings of the Blohm & Voss BV 222 Wiking

Blohm & Voss BV 222 V1 to V6 and V8 (A Series)

Blohm & Voss BV 222 V1, Werknummer 222/365, initially wore the civilian registration D-ANTE, equipped with Bramo 323 R Fafnir radial engines, first flight on 7 September 1940. Code CC+EQ from July 1941, on 10 May 1942 attached to Lufttransportstaffel-See 222 as X4+AH. During a night landing in February 1943 collided with a wrecked ship just under the surface in the port of Athens and sank. Later scrapped.

Blohm & Voss BV 222 V2, Werknummer 222/366, CC+ER, equipped with Bramo 323 R2 Fafnir radial engines, first flight on 7 August 1941. Attached to Lufttransportstaffel-See 222 as X4+BH on 10 August 1942. From May 1943 it served as a long-range reconnaissance aircraft under the Fliegerführer Atlantik. When the war ended it was captured by British troops in Norway. It was subsequently turned over to the US Naval Test Division, test flown and finally blown up in Trondheim.

Blohm & Voss BV 222 V3, Werknummer 222/439, DM+SD, equipped with Bramo 323 R2 Fafnir radial engines, first flight on 28 November 1941. On 9 December 1941 attached to Lufttransportstaffel-See 222 as X4+CH. From May 1943 it served as a long-range reconnaissance aircraft under the Fliegerführer Atlantik. On 20 June 1943 it was sunk in Lake Bicarosse, together with the BV 222 V5, in an attack by DeHavilland Mosquito IIs of No. 264 Squadron, RAF.

Blohm & Voss BV 222 V4, Werknummer 222/000 0004, DM+SE, equipped with Bramo 323 R2 Fafnir engines, first flight on 9 April 1942, on 20 April 1942 attached to Lufttransportstaffel-See 222 as X4+DH. From May 1943 it served as a long-range reconnaissance aircraft under the Fliegerführer Atlantik. When the war ended it was blown up by its crew in Kiel-Holtenau.

Blohm & Voss BV 222 V5, Werknummer 222/000 0005, equipped with Bramo 323 R2 Fafnir engines. First flight on 3 July 1942, on 7 July 1942 attached to Lufttransportstaffel-See 222. From May 1943 it served as a long-range reconnaissance aircraft under the Fliegerführer Atlantik. On 20 June 1943 it was sunk in Lake Bicarosse, together with the BV 222 V3, in an attack by DeHavilland Mosquito IIs of No. 264 Squadron, RAF.

Blohm & Voss BV 222 V6, Werknummer 222/000 0006, equipped with Bramo 323 R2 Fafnir engines. First flight on 19 August 1942, on 21 August 1942 issued to Lufttransportstaffel-See 222 as X4+FH. On 24 November 1942 it was shot down by a Bristol Beaufighter of No. 272 Squadron, RAF near the island of Pantelleria.

Blohm & Voss BV 222 V8, Werknummer 222/000 0008, equipped with Bramo 323 R2 Fafnir engines, first flight on 20 October 1942. Issued to Lufttransportstaffel-See 222 as X4+HH on 26 October 1942. On 10 December 1942 it was shot down by a Bristol Beaufighter of No. 227 Squadron, RAF south of Malta.

B Series

The Blohm & Voss BV 222 B civilian flying boat was a project that was not brought to fruition on account of the war.

Blohm & Voss BV 222 C "Wiking"

Blohm & Voss BV 222 V7, first prototype of the C series, was also designated C-07, Werknummer 222/031 0007, equipped with Junkers Jumo 207 C diesel engines, first flight on 14 April 1943 as TB+QL. From 16 August 1943 it served as a long-range reconnaissance aircraft under the Fliegerführer Atlantik. When the war ended it was blown up by its crew near Travemünde.

Blohm & Voss BV 222 V9, first C series pre-production aircraft, later designated C-09, Werknummer 222/031 0009, from 23 July 1943 served as a long-range reconnaissance aircraft under the Fliegerführer Atlantik as TB+QM. When the war ended it was on land in Travemünde with its engines removed and was captured there by Allied troops.

Blohm & Voss BV 222 V10, later designated C-010, Werknummer 222/031 0010, equipped with Junkers Jumo 207 C diesel engines, first flight on 17 July 1943. Served as a long-range reconnaissance aircraft under the Fliegerführer Atlantik with the code TB+QN. On 8 February 1944 it was shot down by a DeHavilland Mosquito of No. 157 Squadron, RAF near Bicarosse.

Blohm & Voss BV 222 V11, later designated C-011, Werknummer 222/33 0051, equipped with Junkers Jumo 207 C diesel engines, first flight on 16 October 1943. Served as a long-range reconnaissance aircraft under the Fliegerführer Atlantik as TB+QO. According to the latest information the V11 was not shipped to the USA, however nothing more is known about the fate of this flying boat.

Blohm & Voss BV 222 V12, later designated C-012, Werknummer 222/33 0052, equipped with Junkers Jumo 207 C diesel engines, first flight on 23 November 1943. Served as a long-range reconnaissance aircraft under the Fliegerführer Atlantik. When the war ended it was captured by British troops in Norway and in 1945 was flown to Great Britain, where it received the serial VP 501 for test flying. The aircraft was scrapped in April 1947.

Blohm & Voss BV 222 V13, later designated C-013, Werknummer 222/33/0053, equipped with Junkers Jumo 207 C diesel engines, first flight on 18 April 1944. It was given the code DL+TY for long-range reconnaissance operations, however it saw no action. According to the latest information the V13 was not shipped to the USA, however nothing more is known about the fate of this flying boat.

Blohm & Voss BV 222 C-014 to C-017 were in various stages of construction when the war ended but were never completed. They were envisaged as prototypes for the D and the later E series.

Blohm & Voss BV 222 C-020 was supposed to have been the first E-series aircraft.

The Blohm & Voss BV 222 Wiking at the End of the War and Its Subsequent Testing by the Allies

The Blohm & Voss BV 222 Wiking at the End of the War and Its Subsequent Testing by the Allies

When the war ended, the Blohm & Voss BV 222 V2 and BV 222 C-012 fell into British hands in Sörreisa near Bardufoss in Norway. Both flying boats were subsequently moved south to Trondheim, where they received British markings. The BV 222 V2 was later passed on to the US Navy and the RAF roundels were replaced by American flags on the forward fuselage. The British had previously removed the Wiking's armament and electronic equipment. Commander H.E. McNelly and Lt.Cmdr. G.M. Hebert of the US Naval Test Division began testing of the BV 222 V2 in Trondheim in August 1945, although this consisted merely of two 45-minute flights. Flugkapitän Hauptmann Möhring of the Erprobungsstelle Travemünde, who had flown more than 250 missions in the V2, and his last crew, which included Leutnant Steinbach, Leutnant Warninghof and Oberfeldwebel Pausinger, assisted the Americans during these test flights. A series of problems, especially with the engines, prevented more extensive trials, however, and the BV 222 V2 was ultimately blown up off Trondheim. The aircraft remains there today, lying in 180 to 250 meters of water. The exact position of the Wiking is not known, however over the years various parts of the aircraft have turned up in the nets of Norwegian fishermen.

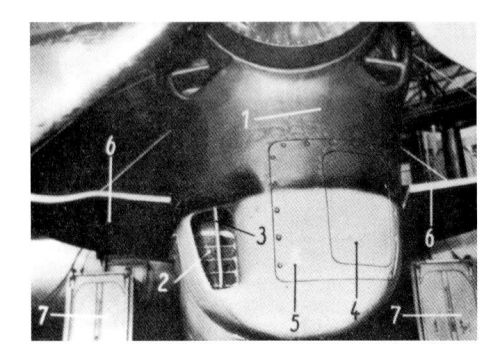

Attached to the Stab of Seeaufklärungs-Gruppe 130 and marked with a letter "R", the Blohm & Voss BV 222 C-012 (Werknummer 222/33 0052, DL+TX) also fell into British hands in Sörreisa. In July 1945 the flying boat was flown from northern Norway to Trondheim. There one of the best known British test pilots, Captain Eric M. "Winkle" Brown, carried out a brief test flight accompanied by German personnel. Because of his experience flying Supermarine Walrus and Sea Otter flying boats, Brown was selected to test fly the captured BV 222 by the Royal Aircraft Establishment (RAE). Brown objected to the assignment, pointing out that the small British flying boats he had piloted were in no way comparable to the huge BV 222. His superiors, however, declared that he was the only RAE test pilot with experience on seaplanes and flying boats and had thus been selected for the job. In July 1945 he and two other officers flew to Norway to test a Wiking in the hands of British forces and subsequently ferry it to Calshot. Brown was able to glean some information about the BV 222 before his first flight, as members of the Blohm & Voss design team had been taken to England immediately after the end of the war and were questioned by RAE technical staff. Several days after flying a Blohm & Voss BV 138 from Schleswig Lake, Brown arrived in Trondheim and saw the BV 222 C-012 lying at anchor and his initial enthusiasm very quickly disappeared. He wrote of his first encounter with the German flying boat: "It rode at its moorings in the fiord like some floating colossus, and its sextet of six-cylinder diesel engines looked dirty, somewhat weary and definitely small to lift this mighty beast from the water." The German crew of the C-012 and the British members of the RAE subsequently conducted a thorough inspection of the flying boat. Capt. Brown was most impressed by the size of the flying boat with its spacious cockpit and armor glass windscreen. His first flight in the Wiking was scheduled for the next day and a Luftwaffe major was assigned to assist him. The weather was fine with light winds on the day of Brown's first flight in the C-012. During initial cockpit checks he discovered that the fuel tanks were only about one-third full. As well, there were no takeoff-assist rockets attached beneath the wings. After taking his position in the right seat, he observed the flying boat casting off from the quay and the starting procedure for the six Junkers Jumo 207 C diesel engines. He was surprised by the good forward visibility from the cockpit. The C-012 was then taxied to the middle of Trondheim Fiord, during which Brown was extremely impressed by the big flying boat's maneuverability on the water. This was helped in no small part by the ability to employ reverse thrust with the aid of the variable-pitch propellers on engines 2 and 5. When the German pilot, a Luftwaffe Major, advanced the BV 222's throttle levers, the acceleration was quite apparent and Brown was forced back into his seat. At that point Brown noticed the German pilot's indifference during the entire takeoff process. The C-012's rate of acceleration slowed in the following seconds. After what seemed like an endless takeoff run, the Wiking finally attained sufficient speed for the tail to be raised and the aircraft to be placed "on the step". Nevertheless, the speed did not appear to be sufficient for takeoff. Capt. Brown was also struck by the unusually loud noise of the water foaming around the flying boat's hull. Eric Brown, who so far had not intervened in the handling of the flying boat, saw that the

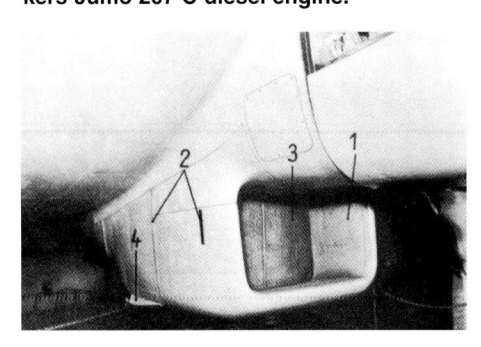

The engine cowling chin panel of the Junkers Jumo 207 C diesel engine.

This photo shows the intake shaft which provided cooling air to the Junkers Jumo 207 C.

Several details of an engine from the BV 222 C: 1 = oil cooler, 2 = Junkers Jumo 207 C diesel engine, 3 = engine cowling chin panel.

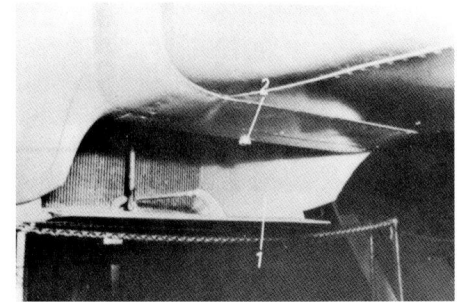

Radiator flap of a Junkers Jumo 207 C engine in the open position.

time had come to abort the takeoff. It later turned out that the Wiking's tail surface controls had been blocked for most of the takeoff and that the German Major had only unlocked them shortly before Brown aborted the takeoff. The German Major in the left pilot's seat, who had boycotted the flying boat's takeoff, probably got an earful from Brown as the C-012 taxied back to the pier in Trondheim. The next day he made a second attempt in the C-012, taking his own two crewmembers and the second available German pilot. This time the BV 222's six engines were running at full power before it "got up on the step" and soon afterwards the Wiking lifted off the water at a speed of 145 kph. It then took about 20 seconds for the two floats to retract into the wings. This process had to be completed before the flying boat reached a speed of 225 kph. Capt. Brown was struck by the lightness of the controls, requiring only a slight application of force by the pilot. The BV 222 was as sluggish in the air as it was maneuverable on the water, a fact that Brown attributed to it being underpowered. The lightly-loaded C-012 reached an altitude of 1640 meters in ten minutes. Brown then leveled off and accelerated the Wiking to 290 kph and then reached 400 kph in a gentle dive. Once again he was struck by the lightness of the big flying boat's controls. On approach to land, he lowered the floats at 200 kph and at 180 kph lowered the electrically-powered flaps to 20°. Just above the water he reduced speed to 120 kph and lowered the flaps to their maximum deflection of 40°. Brown then set the flying boat down on the surface of the water at a speed of 130 kph. In his final report, Brown wrote of the Blohm & Voss BV 222 C-012: "The BV 222 is certainly a remarkable aircraft, not just because of its size, but also because of its unusual control system which, despite its effectiveness, is dangerously effortless for such a huge machine …".

Although Brown judged it rather negatively, the technicians from the Royal Aircraft Establishment were very interested in the BV 222's innovative control system developed by Dr.-Ing. Richard Vogt. It was later installed in an Avro Lancaster II, which was subsequently also flown by Capt. Eric M. Brown.

On 14 July 1945 the C-012 was flown to Copenhagen with just five functioning engines and the next day made a local test flight over Kastrup Lake. On 16 July the BV 222 flew to Travemünde and Sylt/Rantum and the next day from there to Calshot one of the most important British flying boat bases. Actual flying time from Trondheim to Calshot was 12 hours and 55 minutes, and there the flying boat was tested for another 1 hour and 45 minutes. The limited flight time in Great Britain was due to the C-012's inherent engine problems. On 24 July and 23 August 1945 other pilots came from the Royal Aircraft Establishment in Farnborough to test fly the BV 222 C-012. On 5 April 1946 the flying boat which had been painted white like British flying boats was given the British serial VP 501 and the letter "R" in addition to the roundels. The new British paint scheme, the serial VP 501 and the letter "R" gave rise to the rumor that the Blohm & Voss BV 222 C-012 had temporarily entered service with No. 201 Squadron of the Royal Air Force, which was stationed at Calshot and operated Sunderland flying boats. There is nothing in British records to confirm this rumor and it appears groundless because there were no spares for the German flying boat and its troublesome engines in Great Britain. Although the application of a single letter behind the British roundel to identify individual flying boats was common toward the end of the Second World War, it appears that the "R" on the fuselage of the C-012 was simply the same letter worn by the aircraft when it was DL+TX attached to the Stab of Seeaufklärungs Gruppe 130.

The C-012 was supposed to be transferred to the Marine Aircraft/Armament Experimental Establishment (MAEE) in Felixstowe for more extensive trials. This did not happen, however, for in addition to the continuing engine problems with the BV 222 C-012 there were also difficulties with the flying boat's docking system. On 30 March 1946, for example, the Wiking was slightly damaged as it was being brought ashore on a slip wagon by members of the MAEE. A similar accident occurred on 21 June 1946 when the flying boat was placed into the water. Mechanics of Short Brothers Ltd. from Rochester repaired the C-012 on the spot and on 1 August 1946 it was put back into the water. By that point, however, two of the Junkers Jumo 207 C engines on the right wing were no longer functional. The Royal Aircraft Establishment therefore decided to cancel all further tests with VP 501 and the Wiking was offered to the British aviation industry for study purposes. Finally, after no interest was shown in the Blohm & Voss BV 222 C-012, the aircraft was scrapped by the 49th Maintenance Unit between April and June 1947. On 11 June 1947 the Wiking with the British serial VP 501 was struck off charge by the Royal Air Force. Its Junkers Jumo 207 C diesel engines were given to the British engine maker D. Napier & Sons in Luton, which was then beginning development of diesel engines for aviation use. The result of the experiment was the Napier Nomand a so-called turbo-compound power plant, a mixture of a diesel engine and a turboprop which appeared in 1949-50.

Towards the end of the war, KG 200 employed two Blohm & Voss BV 222s for special operations.

Contrary to what has been printed elsewhere, the Blohm & Voss BV 222 C-011 and C-013 flying boats were not sent to the USA after the Second World War. The persistent engine problems encountered by the British in trials with the BV 222 C-012 caused the US Naval Test Division to change its mind about ferrying captured flying boats of this type to the United States. Nothing more is known about the true fate of these two Wikings, however. The Blohm & Voss BV 222 V2, V4 and V7, which were still operational, were blown up by their crews at the end of the war. According to various sources the BV 222 V9 or C-09 was badly damaged in attack on Travemünde by Hawker Typhoon or P-51 Mustang fighter-bombers and was subsequently also destroyed by its crew. This claim is contradicted by a photo taken in Travemünde shortly after the end of the way, showing the aircraft parked on land with its engines removed.

Little is known about the two BV 222s that served with Kampfgeschwader 200, which was formed on 20 February 1944. The two BV 222s may have been the C-011 and C-013. The following can be said to back up this speculation: first, it is known that all Wikings that were not destroyed during the war or fell into Allied hands in May 1945 (V2, C-09 and C-012) were destroyed by their crews at war's end in Kiel-Holtenau, Travemünde and Trondheim (V4 and V7). If these flying boats included the C-011 and C-013, there is no discernable reason why as in the case of the other BV 222stheir capture by the Allies or destruction by their crews, which should have taken place at about the same time, would not be common knowledge. Second, there are clues that two Wikings were in service with KG 200 at the end of the war and were probably stationed in Flensburg. And finally, nothing is known about BV 222s in Flensburg captured by the Allies or destroyed by their own crews. In the final weeks of the war KG 200 undertook special operations exclusively, using a smorgasbord of different aircraft. In addition to more than 30 types of German aircraft including the Arado Ar 232, Dornier Do 217, Focke-Wulf Fw 190, Junkers Ju 188 and Junkers Ju 290the unit operated the Italian Savoia-Marchetti S.M.79 Sparviero and S.M.82 Canguro plus captured Allied types like the Amiot 143, Boeing B-17 Flying Fortress and Consolidated B-24 Liberator the list included two Blohm & Voss BV 222 Wikings. According to the relevant literature, however, these two flying boats do not appear to have been officially attached to KG 200 and were probably given to the unit "on loan". In the spring of 1945 3./KG 200 was based in Rügen with a variety of marine aircraft. Towards the end of the war the unit was moved to Flensburg, about 260 km to the northeast. The Staffel's main mission was depositing secret agents, primarily on the British coast. According to unconfirmed reports one of the last missions given the BV 222s of 3./KG 200 was to transport leading figures of the German government to a remote Norwegian fiord together with supplies sufficient to keep them alive for months. The idea was not carried out, however, and other versions of the story speak of members of the government being flown to Spain in Junkers Ju 290s or to Japan in six-engined Ju 390s.

In August 1943 work on the BV 222 C-014 to C-017 was halted because of the so-called "Emergency Fighter Program". There were plans to develop these into the subsequent D and E versions. The BV 222 D was to have been powered by Junkers Jumo 207 diesel engines, but the power plant was still not operationally mature when the war ended. It was planned that the BV 222 C-014 to C-017 would be the prototypes of the new D-0 series. The Blohm & Voss BV 222 C-020 was to have been the first E-series production aircraft.

Finally, a small mystery for interested readers concerning the Blohm & Voss BV 222 V3, V7 and other aircraft of the C-series. There is a photo of a Wiking equipped with Junkers Jumo 207 C engines in which the partial code ?? + CH may be seen clearly. Based on current knowledge, this marking only fits the BV 222 V3, which bore the code X4+CH, but which was powered by Bramo 323 R2 Fafnir radial engines. The fact that the Jumo 207 C engines visible in the photo were only used by the C series and its prototype the V7, while the recognizable part of the code only matches that of the BV 222 V3, gives rise to the question whether the V3's Bramo engines might have been replaced by Junkers diesel engines. Another, but even more difficult to justify answer for this contradiction would be that the code was exchanged between the flying boats. In this case the question as to why the code was exchanged would raise another puzzle.

As one can see, the story of the Blohm & Voss BV 222 Wiking, one of the most elegant flying boats of the 1940s, is far from being fully explored and continues to offer interested parties room for speculation and new information.

The Blohm & Voss BV 222 C was employed as a strategic reconnaissance aircraft and an armed transport.

This Wiking still represents an unsolved mystery. Based on its code X4+CH it should be the BV 222 V3 (A series), …

…however it has the Junkers Jumo 207 C engines of a BV 222 C.

Technology and Construction of the Blohm & Voss BV 222 C Wiking

Wings

The wings of the Blohm & Voss BV 222 C were of all-metal cantilever construction. The wing was built in three parts and the rectangular section, which was permanently attached to the fuselage, had a profile thickness of 20% over a span of 25.20 meters. The entire wingspan of the flying boat, on the other hand, was 46 meters and the wing area was 255 m². The wing chord was 6.03 meters and dihedral was 2º. The trapezoidal outer wings, each with a span of 10.40 meters, tapered in chord from the previously cited 6.03 meters to 4.05 meters and the dihedral of the outer wings changed to 4º. The wing's aspect ratio was 8.3:1 and the wingtips were 5.2 meters higher than the hull's keel. On Rib No. 1 of each outer wing there was an extendable cable with a loop. These were used to tie up the flying boat when at anchor or to steer the BV 222 during towing. A 33-meter-long tubular spar, consisting of stepped sections of sheet steel welded together, made up the load-bearing wing component and passed through the flying boat's fuselage just above the upper floor. The spar had a maximum diameter of one meter and the sections in the outer wings were made of Dural. This is the trade name for Duralumin, a light alloy of aluminum, copper and magnesium characterized by a high level of hardness, toughness and corrosion resistance. The thickness of the spar walls and its diameter decreased toward the outer tip. The tubular spar was also used to house the K1 diesel aircraft fuel, and following removal of sealing plates held in place by screws it could be entered for the purpose of interior checks. The spar contained six tanks, each with a capacity of 2880 liters. The rear spaces of the booms welded to the inner spar served as oil reservoirs for the power plants. Also located in the spar area was another divided space which served as a container for the propeller de-icing fluid. The flying boat also had a fuel jettison system. The crawlways in the wings, located fore and aft of the spar, not only provided access to the power-operated turrets in the wings and the propulsion system but were also used for pressure fueling. To prevent the flying boat from capsizing during fueling, attention had to be paid to filling the tanks in the proper sequence. The fuel tanks were filled in the following order: 1, 6, 2, 5, 3 and 4. Attached to the tubular spar were the rib connection flaps, the engine mount substructures, the heating and jacking fittings, the screw joints for the engine fittings and attachment points for the "R-Geräte". The so-called "R-Geräte" or "Rauch-Geräte", were takeoff-assist rockets, whose thrust helped the flying boat overcome water resistance during takeoff. The rockets were jettisoned after takeoff and lowered to the ground by parachute. On the aft edge of the wing center section there were electrically-operated, track-guided landing flaps. Their maximum deflection angle was restricted to 30º for takeoff and 41º for landing. Two-part ailerons formed the trailing edges of the outer wings. The outer ailerons were trimmed by spindle drive. Given the size of the Blohm & Voss BV 222, conventional controls would have required too much physical force from the crew. Control forces were significantly reduced by an arrangement whereby the pilot only directly controlled the smaller outer ailerons, while the larger inner ailerons were controlled by auxiliary tabs (Flettner tabs) and the electrically-actuated slotted flaps by the Fowler System. Developed by Dr.-Ing. Richard Vogt, this effective control system was the most important innovation of the Blohm & Voss BV 222 Wiking and after the war it attracted a great deal of interest from Allied technicians.

Stabilizing Floats

The two split stabilizing floats of the Blohm & Voss BV 222 were attached to one of the sturdy wing ribs and stabilized the flying boat on the water. Each was located 15.75 meters from the aircraft centerline. When the flying boat was in a level position, the two floats were exactly one meter above the surface of the water. The floats were retracted into the wing undersides electrically by means of cables. The split floats folded left and right and, halved in this way, retracted into recesses in the underside of

When the war ended, the BV 222 V2 was captured by British forces in Sörreisa near Bardufoss. It was subsequently flown south to Bardufoss, where it was given British markings.

the wing. To lower the floats a safety catch was released and the floats came down under their own weight and aerodynamic forces on their streamlined fairings.

Fuselage

The all-metal central "boat hull" of the Blohm & Voss BV 222 C with its five "auxiliary steps" aft of the "main step" and its total of 63 frames had a total length of 37 meters, 27 of which were taken up by the underwater keel space, which was divided by twelve bulkhead frames. The part of a flying boat called the "boat hull", "boat fuselage" or just "boat" is the hydrodynamically-designed bottom of the flying boat's fuselage, which makes it possible for the aircraft to take off from the water and is optimized to absorb impact loads on landing. This "boat fuselage" was a so-called "two step boat" of monocoque construction and consisted of corrosion-resistant light metal sheet with skin thicknesses between 3 and 6 mm. A "step" ("cross step") was a step-shaped transverse section located just behind the flying boat's center of gravity. It helped get the flying boat off the water during takeoff, and the formation of a trough just behind the step also helped reduce water resistance. The forecastle was 15.5 meters long, while the section aft of the step measured 11.5 meters. The "boat" of the Blohm & Voss BV 222 C was 5.67 meters high, 3.08 meters wide and had a depth of 1.45 meters. With several exceptions, the frames were made of open profiles, mostly fabricated from folded sheet and web plates. The aforementioned exceptions affected Frames 27 and 30 located behind the wing spar, which were reinforced, and Frames 39, 44 and 55, which were made as full bulkheads with passageways. The lower deck was used mainly as a cargo compartment, which could comfortably be accessed by way of the large sideways-folding cargo hatch (nose cone to Frame 7). Behind this cargo compartment (Frames 7 to 39), in the rear of the lower deck, there was a lounge for the crew (Frames 39 to 44) and behind that the tail section (Frame 44 to the "hull tail").

On the upper deck in the fuselage of the Blohm & Voss BV 222 was the cockpit (Frames 7 to 22). In the cockpit there were dual controls and from these control rods ran beneath the upper deck floor along the left side of the fuselage, subsequently continuing via various redirections to the control surfaces in the wings and tail section. Behind the cockpit was the utility room (Frames 22 to 26), followed by the spar room (Frames 26 to 33) and a baggage room (Frames 33 to 44), which was also the end of the upper deck. The rear part of the BV 222 C was accessible by means of a catwalk to the empennage attachment point.

The outer skin of the flying boat was flush-riveted sheet metal. There were fittings on the nose and tail for tying up the flying boat. A towing system in the Wiking's nose was rated for 10000 kg and the slip hook attached to Frame 55 with a maximum allowable load of 5700 kg was activated from the cockpit.

Empennage

As with the wing of the Blohm & Voss BV 222 C, tubular spars formed the load-bearing elements of the cantilever fin and tailplane. The rudder of the 22.90-m² vertical tail was actuated by two servo tabs. With a span of 15 meters and an area of 39.6 m², the horizontal tail was situated at a height of 5.75 meters measured from the keel of the "boat hull" and was built as follows. Each elevator consisted of an inner section with an auxiliary tab, a center section which served as a stabilizing surface and an outer section which was a mass-balanced landing flap. The ailerons, elevators and rudder could be locked electrically by means of spring bolts. A space behind the spar of the vertical tail, which served as a tail sea stand, could be reached by way of a passageway in Frame No. 55.

Propulsion System

The Blohm & Voss BV 222 C's propulsion system consisted of six liquid-cooled, 6-cylinder, two-stroke, twin-shaft Junkers Jumo 207 C diesel engines which were mounted on the inner wing. The Wiking's six engines were interchangeable and each had its own fuel and oil system. Designed as a high-altitude engine, the Jumo 207 C had an exhaust-driven supercharger for a climbing performance altitude of 5000 meters and produced 1,000 hp (735 kW) for takeoff at 3,000 rpm. The Wiking's engine mounts were innovative tubular structures. The Jumo 207 C was attached at four points, with two fixed bearings on the left and spring bearings on the right. They could thus swing parallel to the axes and as the engine cowlings were attached directly to the engines, these swung with the engines.

A 170-liter oil tank and a cooling system were attached to each engine. The radiators were mounted on frameworks beneath the wings. The oil tanks could be topped up in flight from a 950-liter tank located in a section of the wing spar within the hull. The coolant used was a water-glycol mixture with a ration of 50:50, to which was added 1.5% Rust Preventative Oil 39. Its tank was also located in the spar room. The quantity of coolant per engine was given as 84 liters and the capacity of each compensating reservoir was about 32 liters. The engines drove VDM variable-pitch, three-blade propellers made by the Schwarz company. Equipped with wooden blades, the propellers had a diameter of 3.3 meters. The propellers fitted to the No. 2 and No. 5 engines could provide reverse thrust by setting the propeller blades.

Equipment

A DKW auxiliary power unit BL 500 provided electric power to the Blohm & Voss BV 222 C's systems. The power unit was started and warmed up using gasoline, after which it was fed diesel fuel from the main tanks. Electronic equipment differed from aircraft to aircraft. The following systems were installed in a variety of configurations: a FuG 200 Hohentwiel maritime search radar, a FuG

When the BV 222 V2 was later handed over to the Americans, the British roundels on the fuselage were replaced by American flags.

16Z voice and telegraphy communications plus homing radio set (in combination with the FuG 1) made by the Lorenz company for radio beam guidance, a FuG 216 Neptun rear-warning radar by FFO, a FuG 101a sensitive altimeter by the Siemens company, a FuG 25a Erstling IFF system by the GEMA company, a Peil G 6 direction finder and a FuBl 2 instrument landing system. In many sources a VP 245 long-wave radio made by Lorenz is given as radio equipment in the BV 222 V2. Like all multi-engine aircraft used by the Luftwaffe, the BV 222 C also had a marine emergency kit, which, weighing about 35-45 kg, consisted mainly of an inflatable lifeboat, an emergency transmitter and food concentrates.

Crew and Armament

The Blohm & Voss BV 222 C was flown by a crew of ten consisting of a pilot, copilot, navigator, first air machinist, second air machinist, first radio operator, second radio operator, and three flight engineers.

In addition to their primary duties, the crew also had to operate the flying boat's defensive armament. This consisted of a WL 131 nose position with a 13-mm MG 131 made by Rheinmetall-Borsig in a roller bearing mount. On the fuselage spine was a HD 151/D hydraulically-powered rotating turret armed with a 20-mm MG 151/20 (EZ) by Mauser (B-1 Stand). In many sources this turret is identified as a DL 151.

Four SL 131 socket mounts were installed in the fuselage sides and these each held a 13-mm MG 131 by Rheinmetall-Borsig. It is claimed that additional

Blohm & Voss BV 222 "Wiking" Specifications

	Blohm & Voss BV 222 C	**Blohm & Voss BV 222 V4**
Propulsion:	six Junkers Jumo 207 C diesel engines each producing 1,000 hp or 735 kW for takeoff	six Bramo 323 R Fafnir nine-cylinder radial engines each producing 1.200 hp or 882 kW for takeoff
Wingspan:	46.00 m	46.00 m
Overall length:	37.00 m	36.50 m
Step width:	3.08 m	3.08 m
Overall height:	10.90 m	10.90 m
Fuselage height:	5.67 m	5.67 m
Height including antenna mast:		
Draught of flying boat with:	7.00 m	7.00 m
30.000 kg payload:	–	1.22 m
40.000 kg payload:	–	1.38 m
45.000 kg payload:	–	1.45 m
Wing area:	255.00 m²	255.00 m²
Wing thickness inboard:	6.03 m	6.03 mt
Wing thickness outboard:	4.05 m	4.05 m
Aspect ratio:	1:8.3	1:8,3
Displacement of the bracing floats:	2.80 m²	2,80 m²
Distance of bracing floats from centerline:	15.75 m	15.75 m
Horizontal tail area:	39.60 m²	39.60 m²
Vertical tail area:	14.80 m	14.80 m
Fin/rudder area:	22.90 m²	22.90 m²
Rudder area:	6.65 m²	6.65 m²
Rudder deflection	± 25°	± 25°
Empty weight:	30.650 kg	28.545 kg
Maximum takeoff weight:	49.000 kg	45.600 kg
Payload:	15.340 kg	18.000 kg
		92 fully-equipped troops[1] or 72 litters
Maximum speed at ground level:	294 km/h	345 km/h bzw. 295 km/h with maximum payload
at 4500 meters:	–	385 km/h
in 5000 meters:	–	390 km/h
Cruise speed		
at ground level:	305 km/h	320 km/h
in 5500 meters:	345 km/h	–
Landing speed:	125 km/h	125 km/h
Takeoff distance:	1200 m	1200 m
Service ceiling:	7300 m	6500–6700 m
Range:	6095 km	7000–7450 km
Armament:	five 13-mm MG 131 in individual mounts and three to four 20-mm MG 151/20 in individual mounts	five 7.92-mm MG 81 in individual mounts and two 13-mm MG 131 in rotating turrets on the spine.

Other weapons configurations were also used.
No air-dropped weapons were carried for use against land or sea targets.

---

[1] According to other sources it was 110.

SL 131 side mounts could be installed if necessary.

The two HD 151/2A turrets mounted on the wings, each armed with a 20-mm MG 151/20 (EZ), were a novelty. The two weapons positions on the wings could be reached by crawlways in the wings. If required, another HD 151/D rotating turret could be installed farther aft on the fuselage spine (B-2 Stand). The fittings for the second dorsal turret on the spine of the BV 222 were completely installed in the C version. Only the B-2 turret itself was not part of the equipment set and was only fitted when needed.

The BV 222 was not capable of carrying dropped loads for attacking land and sea targets. Several BV 222 Cs were fitted with the earlier 15-mm MG 151 in place of the 20-mm MG 151/20 (EZ).

Below: After its capture, the Blohm & Voss BV 222 C-012, which last saw service with Seeaufklärungs Gruppe 130, was test flown by personnel of the Royal Aircraft Establishment and subsequently flown to Calshot in England.

Above: At the British seaplane base in Calshot the BV 222 C-012 was painted in the colors worn by British flying boats and received the new serial VP 501. The letter "R" next to the roundel was adopted from Seeaufklärungs Gruppe 130, however in many circles it led to the scarcely believable speculation that the C-012 was placed into service for a short time by the Royal Air Force.

Conclusion

I would like to thank Hans-Jürgen Becker and Walter Strobl for their help and support in producing this monograph. I also wish to thank Peter P.K. Herrendorf, Prof. Wilhelm Hesz and Friedrich Müller for their help with the photographs.

Rudolf Höfling
Wien, February 2003

In Europe, Schiffer books are distributed by:
Bushwood Books
6 Marksbury Avenue
Kew Gardens
Surrey TW9 4JF, England
Phone: 44 (0) 20 8392-8585
FAX: 44 (0) 20 8392-9876
E-mail: Info@bushwoodbooks.co.uk
Visit our website at: www.bushwoodbooks.co.uk

Published by Schiffer Publishing Ltd.
4880 Lower Valley Road
Atglen, PA 19310
Phone: (610) 593-1777
FAX: (610) 593-2002
E-mail: Info@schifferbooks.com.
Visit our web site at: www.schifferbooks.com
Please write for a free catalog.
This book may be purchased from the publisher.
Try your bookstore first.

Copyright © 2012 by Schiffer Publishing.
Library of Congress Control Number: 2011945955

This book was originally published in German under the title FLUGZEUG Profile 40, Blohm & Voss BV 222, "Wiking"

All rights reserved. No part of this work may be reproduced or used in any forms or by any means – graphic, electronic or mechanical, including photocopying or information storage and retrieval systems – without written permission from the copyright holder.

Book Translation by David Johnston.
Book Design by Stephanie Daugherty.
Printed in China.
ISBN: 978-0-7643-4049-9
We are interested in hearing from authors with book ideas on related topics.

Water and clouds – the BV 222 V2 in its element.

Blohm & Voss BV 222 V2 WNr 222/366
1:72 Model von Revell

GERD BUSSE

(SIM Stuttgart, D/CDN Maple Leaf Modelers Bühl and 1. PMCN Nuremberg)

Die BV 222 V2 mit der Werknummer 222/ The BV 222 V2 (Werknummer 222/366, initial code CC+ER, later X4+BH) entered Luftwaffe service with the Lufttransportführer See 222 (LTS See 222) on 10 August 1942. It is the subject of a Revell kit which will be described for the benefit of the reader who would like to build a model of the BV 222 V2.

Like the original, the model is one of the largest in any collection. The amount of paint needed for the model is equally large: an entire bottle each of Gunze acrylic RLM 02 and RLM 65 was used in airbrushing the model. The marine camouflage colors RLM 72 and 73 were created by adding gray and blue tones to the land camouflage colors (RLM 70/71).

The building notes offer the choice of two camouflage schemes: the standard scheme and a more contrast rich winter scheme, which the V2 wore for "Operation Schatzgräber" in the Arctic. As there is just one known photo of the latter (left side from in front), the model was finished in the standard scheme that it wore with LTS 222 in 1943.

The kit consists of about 250 cleanly-molded parts with fine engraved detail and excellent fit, plus a decal sheet with details easily readable under a powerful magnifying glass.

This article does not deal with all the details of assembly, as these would fill an entire article in a modeling magazine. Instead, with the aid of photos it will show how the impressive original would have looked in color.

Fuselage
The excellent fit of all parts is immediately noticeable. There are many details on the fuselage frames, but they extend too far into the cargo compartment. On the original only Frames 39, 44 and 55 were made as full bulkheads with passageways. Otherwise the interior of the model's fuselage is bare. This simplification (which ensures a smooth exterior surface during manufacture) cannot be seen from outside and, because of the dark interior color (RLM 02), is scarcely noticeable even if the doors are modeled open.

The cockpit and crew compartment behind it are very well detailed. After the fuselage is assembled all of this disappears, unless one makes the fuselage spine in this area removable. In any case the kit provides this option. Color drawings are provided to assist in painting the instrument panel [3].

Everything to do with the engines is colored yellow, with brown for lubricants, blue for air and red for everything that needed to be seen quickly in an emergency.

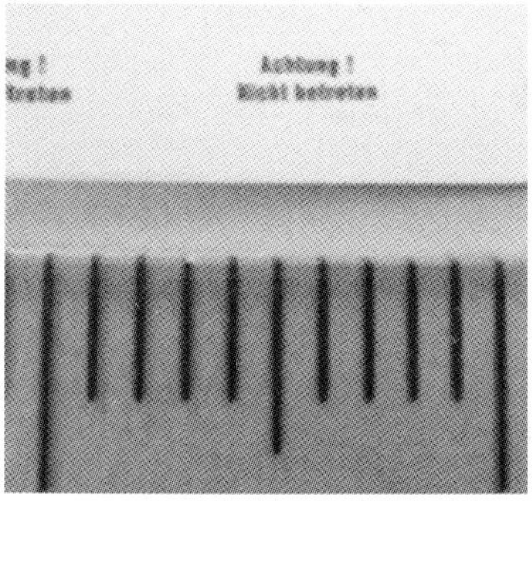

An example of the clarity of the stenciling on the decal sheet.

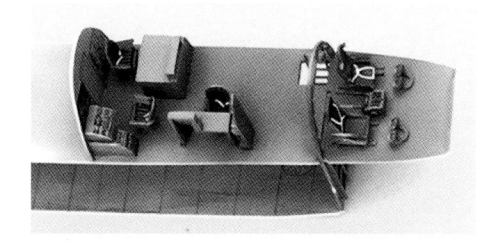

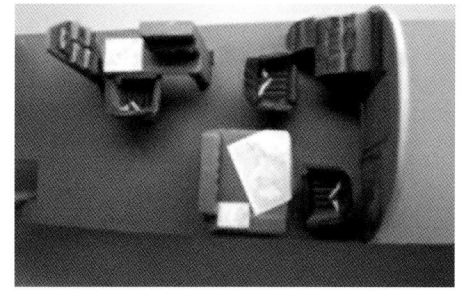

Work stations behind the cockpit – the seemingly open area was occupied by machine-gun positions.t

LUFTWAFFE AIRCRAFT Profile

Interior detail after installation in one of the fuselage halves.

The lead in the nose section ensures that the model has the correct center of gravity.

Numerous eyelets for tying up the flying boat are represented on the fuselage sides just above the waterline. These can be drilled out with a 0.6 mm bit for a more realistic appearance.

The fuselage nose beneath the floor should be filled with lead so that the model's center of gravity is in the forward third of the wing as on the original. If this is not done the model will rest on its tail. The center of gravity is also important if one wishes to place it on the docking wagon (not provided).

The fuselage halves were completed before final assembly and painting and the subsequent installation of the windows. Weathering was achieved with watercolors (umber or black). After it had dried it was rubbed off with a damp rag in the direction of the airflow. Raised features were then emphasized with Lukas rub-on metalizer (Silber 5213).

The official camouflage scheme with dimensions is well-known [1] and confirmed by photos (for example of the fuselage spine of the V1 taken from the angle of the crane operator [4]). It differs in the area of the fuselage spine and sides from the paint scheme in the instructions. Only after final painting are the finely-detailed decals applied, sealed and finally sanded smooth with 1500-grit sandpaper under cold water. Then the windows are installed. The machine-gun positions are then glued in place (MG 81s with Extratec EX 72121 ring sights. These should only be applied after final assembly, otherwise the delicate structures will break off). There were problems with the spent casing chutes of the MG 81s (Parts 32 and 34), which can be eliminated by shortening the center section. After the fuselage halves were glued together the paint in the area of the joints was touched up.

The FuG 200 "Hohentwiel" (which was replaced by a photo-etch part) was also not attached until final assembly, otherwise it will be lost.

Empennage

The two different-size "horns" on the right side of the vertical tail, which connect the fin and rudder, are mountings in which the rudder is mounted. Sufficient space must be available if the rudder is deflected. The recesses were therefore expanded into rectangular cutouts, the rudder separated and fitted with bearing pins so that it can be deflected. As the mounts do not rest on the outer surface of the rudder, but instead extend to half their depth, they were lengthened, causing them to disappear into the rectangles.

The arrangement of the elevator mounts was corrected and detailed like that of the rudder.

The DF loop's rectangular fairing is surely not true to the original on account of drag. The fairing was therefore replaced by a streamlined pedestal.

Wing

The model's airfoil is represented symmetrically. This is actually an aerobatic airfoil which is inappropriate for the BV 222. A photo of the wing structure [3] clearly shows a flat underside. The sharp curvature results in an equally inaccurate shape of the interior sides of the floats and results in too great a distance between the extended floats. There should only be a small gap on the back side.

The shape of the airfoil is difficult to correct, but the floats should be sanded flatter in the extended position.

The weapons positions in the wings interfere with the airbrush work. If one widens the holes under Parts 114 to allow Part 80 to pass through, then the turrets can be installed after the wings have been assembled and painted.

The downwards-folding work stages on both sides of the engines in the original would collide with the exhaust pipes. This should be corrected in case someone wishes to build a diorama depicting work on the engines. Scribing for the inner tank on the starboard wing is missing and should be dealt with.

The wings were completely assembled prior to final assembly, apart from the FuG 101 sensitive altimeter (Part 118), which worked with frequency sweep and evaluation of the beat frequency, and the FuBl 2 instrument landing antenna (Part 119). These are better made by the modeler himself due to molding limitations.

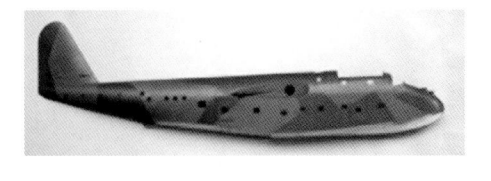

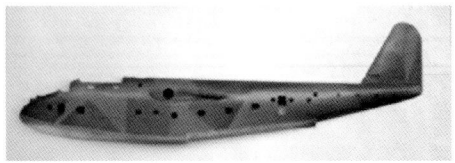

Fuselage: standard camouflage scheme [1].

Modified DF loop mount.

MG 151/20. Barrel by Schatton-Modellbau.

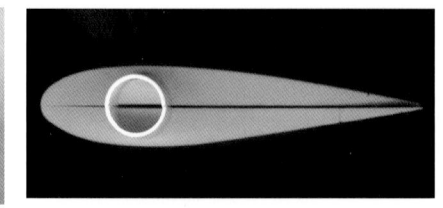

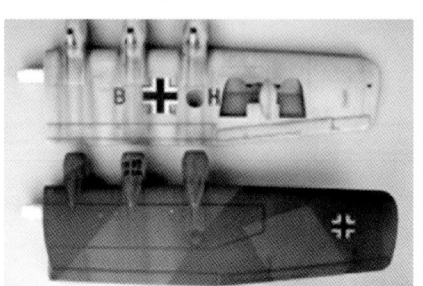

The wing's airfoil section is too curved on the underside.

The wings after painting and weathering prior to installation of the machine-gun position.